DATABASES & ON-LINE DATA IN ASTRONOMY

ASTROPHYSICS AND SPACE SCIENCE LIBRARY

A SERIES OF BOOKS ON THE RECENT DEVELOPMENTS
OF SPACE SCIENCE AND OF GENERAL GEOPHYSICS AND ASTROPHYSICS
PUBLISHED IN CONNECTION WITH THE JOURNAL
SPACE SCIENCE REVIEWS

Editorial Board

R. L. F. BOYD, *University College, London, England*

W. B. BURTON, *Sterrewacht, Leiden, The Netherlands*

C. DE JAGER, *University of Utrecht, The Netherlands*

J. KLECZEK, *Czechoslovak Academy of Sciences, Ondřejov, Czechoslovakia*

Z. KOPAL, *University of Manchester, England*

R. LÜST, *Max-Planck-Institut für Meteorologie, Hamburg, Germany*

L. I. SEDOV, *Academy of Sciences of the U.S.S.R., Moscow, U.S.S.R.*

Z. ŠVESTKA, *Laboratory for Space Research, Utrecht, The Netherlands*

INSTRUMENTATION
VOLUME 171

DATABASES & ON-LINE DATA IN ASTRONOMY

Edited by

MIGUEL A. ALBRECHT
ESO, Garching bei München, Germany

and

DANIEL EGRET
*Observatoire Astronomique
Université Louis Pasteur
Strasbourg, France*

KLUWER ACADEMIC PUBLISHERS
DORDRECHT / BOSTON / LONDON

Library of Congress Cataloging-in-Publication Data

```
Databases and on-line data in astronomy / edited by Miguel A.
  Albrecht, Daniel Egret.
      p.   cm. -- (Astrophysics and space science library ; v. 171)
   Includes index.
   ISBN 0-7923-1247-3
   1. Astronomy--Data bases.  2. Data base searching.  3. On-line
data processing.   I. Albrecht, Miguel A.  II. Egret, D. (Daniel)
III. Series.
QB51.3.E43D38  1991
025.06'52--dc20                                              91-14440
                                                                 CIP
```

ISBN 0-7923-1247-3

Published by Kluwer Academic Publishers,
P.O. Box 17, 3300 AA Dordrecht, The Netherlands.

Kluwer Academic Publishers incorporates
the publishing programmes of
D. Reidel, Martinus Nijhoff, Dr W. Junk and MTP Press.

Sold and distributed in the U.S.A. and Canada
by Kluwer Academic Publishers,
101 Philip Drive, Norwell, MA 02061, U.S.A.

In all other countries, sold and distributed
by Kluwer Academic Publishers Group,
P.O. Box 322, 3300 AH Dordrecht, The Netherlands.

Printed on acid-free paper

All Rights Reserved
© 1991 Kluwer Academic Publishers
No part of the material protected by this copyright notice may be reproduced or
utilized in any form or by any means, electronic or mechanical,
including photocopying, recording or by any information storage and
retrieval system, without written permission from the copyright owner.

Printed in the Netherlands

Table of Contents

Preface		xi
1	**Data from the ROSAT Observatory**	**1**
	1–1 Introduction	1
	1–2 The Instruments	2
	1–3 Satellite Data Flow	2
	1–4 The ROSAT Scientific Data Centre	3
	1–5 Standard Data Processing	4
	1–6 Extended Data Evaluation with EXSAS	5
	1–7 Archiving and Retrieval of ROSAT data at the RSDC	5
	1–8 WFC Data	8
	1–9 Archive Access and Data Rights	8
2	**The EXOSAT Database System**	**11**
	2–1 Introduction	11
	2–2 A Data Archive Infra-structure	12
	2–3 The User Interface	13
	2–4 The Future	14
	2–5 Conclusions	16
3	**IRAS**	**17**
	3–1 Introduction	17
	3–2 Data products	20
	3–3 Data access	26
	3–4 Future developments	26
4	**Databases from Cosmic Background Explorer (COBE)**	**29**
	4–1 The COBE Mission	29
	4–2 Contents of the Database	31

	4-3 Organization of the Data	33
	4-4 Future Plans	34
5	**The many faces of the Archive of the International Ultraviolet Explorer satellite**	**35**
	5-1 Introduction	35
	5-2 The IUE project	35
	5-3 The IUE Archives	36
6	**Data Archive Systems for the Hubble Space Telescope**	**47**
	6-1 Introduction	47
	6-2 Critical design issues for the HST Archive System	49
	6-3 The Capabilities Required of the HST Archive System	52
	6-4 The Data Management Facility (DMF)	53
7	**Database Aspects of the Guide Star Catalog**	**59**
	7-1 Introduction	59
	7-2 History	59
	7-3 Catalog Construction	60
	7-4 Production Data	61
	7-5 GSC Structure and Content	63
	7-6 Enhancements and Plans	64
8	**The HIPPARCOS INCA Database**	**67**
	8-1 The HIPPARCOS mission: requirement for an INCA Database	67
	8-2 The INCA database: creation	69
	8-3 The INCA database: stellar content	70
	8-4 The INCA database: data content	73
	8-5 Usage of the INCA database	74
	8-6 Database software and organization of the data	75
	8-7 Access mode	75
	8-8 Future plans	75
9	**The SIMBAD astronomical database**	**79**
	9-1 Introduction	79
	9-2 The Strasbourg Data Center (CDS)	79
	9-3 The astronomical contents	80

9–4	Retrieving data: three query modes	81
9–5	SIMBAD III: a new Database Management System	82
9–6	How to obtain access to SIMBAD?	84
9–7	The users of SIMBAD	85
9–8	Updating SIMBAD: a very large cooperative effort	85
9–9	Future trends	86

10 The NASA/IPAC Extragalactic Database — 89

10–1	Introduction	89
10–2	Context	89
10–3	Data contents	91
10–4	Functions	99
10–5	Interface	101
10–6	Architecture	103
10–7	Future plans	104

11 The ESO Archive Project — 107

11–1	Introduction	107
11–2	Raw versus Calibrated data	107
11–3	Archiving at ESO	108
11–4	Data Format	109
11–5	The Archive Medium	110
11–6	The Archiving Process	111
11–7	The Catalogue of Observations	112
11–8	STARCAT: The user interface	112

12 Archives of the Isaac Newton Group, La Palma and Westerbork observatories — 115

12–1	Introduction	115
12–2	Contents and organization of the archives	116
12–3	Accessing the La Palma and Westerbork Observations Catalogues	119
12–4	User Profiles of the archives	122
12–5	Future plans	123

13 Archiving at NRAO's VLA and VLBA Telescopes — 125

14 ESIS A Science Information System — 127
14–1 Introduction — 127
14–2 User needs — 128
14–3 The pilot project — 128
14–4 System architecture — 130
14–5 Science Oriented Information Retrieval — 131
14–6 Conclusions — 137

15 The NASA Astrophysics Data System — 139
15–1 Introduction — 139
15–2 System Development — 140
15–3 System Design — 141
15–4 System Operations — 146
15–5 System Usage — 148
15–6 ADS Service Upgrades and Additions — 149
15–7 Further Information — 149

16 The NSSDC Services — 151
16–1 Introduction — 151
16–2 Current services — 152
16–3 Future services — 156
16–4 How to access the NSSDC archives — 158

17 The Space Data Centre at RAL — 161
17–1 Introduction — 161
17–2 Facilities at the SDC — 162
17–3 Data available or expected to be available — 162
17–4 User services — 164
17–5 Future plans — 164

18 Database applications in Starlink — 165
18–1 Introduction — 165
18–2 SCAR — 168
18–3 Other database systems — 174
18–4 Comparison with HDS — 175
18–5 Further reading — 176

19 Database applications in Astronet — 179
- 19–1 Foreword — 179
- 19–2 DIRA: Background — 180
- 19–3 DIRA structure — 182
- 19–4 Personal database — 185
- 19–5 Graphic tasks — 187
- 19–6 DIRA master database content — 188
- 19–7 General characteristic of DBIMA — 189
- 19–8 Retrieving data from archives — 190

20 Database Services at the Canadian Astronomy Data Centre — 193
- 20–1 Context — 193
- 20–2 Databases and Services — 195
- 20–3 Access Mode — 196
- 20–4 User Profile — 197
- 20–5 Future Plans — 197

21 Astronomical bibliography from commercial databases — 199
- 21–1 Introduction — 199
- 21–2 Databases — 200
- 21–3 Vendors — 208

22 Astronomical Directories — 211
- 22–1 Introduction — 211
- 22–2 First compilations — 211
- 22–3 The virus strikes back — 212
- 22–4 Present situation — 215
- 22–5 A question of quality — 216
- 22–6 Speaking of evolution — 219
- 22–7 Other directories — 220
- 22–8 Future trends — 222

23 Science networks: A short overview — 225
- 23–1 Introduction — 225
- 23–2 Wide-area networks — 225
- 23–3 Organisations — 228

    23–4 The future 230

24 User Interfaces in Astronomy   **235**

    24–1 Introduction 235

    24–2 Current Status 236

    24–3 Designing the User Interface Software 243

25 The FITS Data Format   **253**

    25–1 Introduction 253

    25–2 General Structure 254

    25–3 The FITS Data Format 254

    25–4 FITS Committees and Standards 256

    25–5 Conclusions 257

List of Acronyms   **259**

Index   **265**

Preface

Astronomical data archiving

Data archiving was for many years the most disregarded aspect of all data systems. Over the past two decades, the increasing efficiency of cameras and photon collecting devices used by astronomers, especially for space–borne astronomy or ground–based surveys, has generated an unprecedented accumulation of data. But only a small part of the data is immediately used in the investigations that was planned for, and eventually published. Many astronomers observing with ground–based telescopes still prefer to perform a new observation rather than searching for possible archive data. In fact, ever evolving technology does encourage this attitude, in the sense that it is indeed preferable to observe an object or phenomena with the latest available telescopes thus to take full advantage of modern capabilities.

Although one might argue that the ratio of the data actually involved in new discoveries to the data collected is just a natural situation in research and feel unconcerned by the fate of old data, there are at least three good reasons not to leave things as such. The first one is that celestial objects do change with time and we seldom know in advance which part of the old data will make sense: nobody would have said five years ago which of the stars observed by Sanduleak was likely to become SN 1987a. The second reason is that new knowledge emerges from confrontation between results from different techniques at different wavelengths and resolutions: striking examples can be found in the recent progresses in the diagnostic of the interstellar medium or the central regions of quasars. This cross–fertilization of astronomical observations is becoming a major aspect of modern astronomy when space opens new windows, and ground–based techniques go deeper. Finally, the large investments involved both in space–borne and modern ground–based telescopes have raised the question of cost-to-benefit ratio in the management of observing facilities. For instance, the number of observation proposals for space–borne sensors, when operated in the so called 'Guest Observer' mode, largely surpasses the *de facto* available observation time of the facility in question. In this sense a coordinated effort to save and organize archived data in such a way that they remain useful on the long term, and to facilitate their dissemination, does suplement efforts for optimally using very expensive instruments.

From archives to integrated systems

The assumption that a part of all observation projects could be conducted with data already obtained in previous observations of the same object led NASA, ESA and SERC to setup an archive of IUE images, and to service archive data requests from their respective science community. This assumption has proved to be extremely accurate as is shown by the usage statistics of the IUE archive (see the IUE chapter on page 35). A new dimension is currently being explored as HST delivers first useful results. Here, the driving *motivs* are both to cope with *very large* quantities of data (of the order of Terabyte/year) as well as to maximize scientific throughput of observations.

This trend towards larger and larger volumes of data makes it extremely difficult for individual researchers to keep up with their specialities, in terms of being aware of new data as well as tracking new ideas. In response to this problem, attempts have been made to build databases of *results* of scientific research. A pioneer in this area has been the Centre de Données astronomiques de Strasbourg with its Set of Identifications, Measurements and Bibliography for Astronomical Data (SIMBAD). Here, one would find all 'published knowledge' on a particular astronomical object, for instance its spectral class (for stars), magnitude, distance, *etc.*, as well as references to articles dealing with the object.

SIMBAD was originally created as the merging of a number of existing catalogues. It is clear, however, that the pure 'merging' of catalogues would bring serious problems regarding for example naming and syntax conventions: the database would have to resolve these problems by providing a value added layer that would 'glue' the single parts (catalogues, bibliographical references, *etc.*) into a system of its own. This consideration led to the development of the database to its current state (see chapter 9) but also to the establishment of the concept of *results database*, i.e. the integration of information from possibly different sources together with an *ad hoc* developed layer that provides the value added service of making each item appear as part of the whole. Other activities have developed, partly independently, into the same direction: the EXOSAT database and the NASA/IPAC Extragalactic Database are two primary examples (see chapters 2 and 10 respectively). The success of the EXOSAT DB, for instance, shows how much it is appreciated by the community, that a project systematically produces results (products) from the original 'raw' data.

Preface   xiii

The next step along this line was to conceive systems that would integrate into one single environment, databases as a whole, plus information services (*e.g.* bulletin boards, e-mail, *etc.*) and end-user owned data collections—so called *personal databases*. The integrating layer would become highly sophisticated, but the return value would promise both capabilities to conduct multi–spectral research as well as a to handle an always larger number of different data sets. These are the ambitious goals of such systems as ADS and ESIS (see chapters 15 and 14).

Seen in this light, it becomes clear that the management of databases has become an established, well recognised need in the astronomical domain. The number of contributions to this book largely underlines this fact.

This book

As with many other ideas, the idea to publish a collection of papers on databases and—more generally, data accessible over telecommunication networks, materialized around a glass of beer. We had been working on the definition of a 'Query Environment' for the ESIS project, and had collected a considerable amount of information on astronomical data, all to be eventually part of the ESIS system. It soon became clear to us that, while the domain was evolving very quickly, with exciting developments towards intelligent systems and huge volumes of data, there was a real need to produce an up-to-date overview of all the existing systems; in an homogeneous way, in a stand-alone envelope. So, we sat down and started making a list of the systems and collections we knew of—that first shot was indeed not very far away from the table of contents of this book.

The first part of the book contains a series of chapters describing astronomical archives (space missions and large ground–based observatories) and databases (SIMBAD, NED), which are —or, in some cases, will soon become— publicly available. The accent has been put on a general description of the systems, giving reference to more detailed documentation for potential users when available. In the second part, information systems currently under development (ESIS in Europe, and ADS in the U.S.), and on-line services (most of them being national efforts) are presented. The last chapters deal with related issues: bibliographical services, astronomical directories, science networks, user interfaces in astronomy, and a brief description of the FITS format of common use in many systems.

A table of acronyms and a general index can be found at the end.

This work includes a series of clear and up-to-date descriptions of available data systems world wide and is presented in a single volume, with a considerable effort of homogeneity; it would not have been possible without the very kind contributions of the authors of the individual chapters, mostly key persons from each project: thanks to their enthusiasm and efficiency.

The present volume does not claim to be exhaustive: in fact it was not possible to include all existing systems, although we did try to cover as wide an overview as possible during a limited time frame (the whole of year 1990). The next years will bring a wealth of new data systems from both space–borne as well as ground–based observatories.

This book should help the user (astronomer, librarian, computer engineer) to trace his way among a wide variety of existing astronomical data services. It is meant to encourage the creation of new databases and data collections and will eventually be a helpful tool for database managers, at a moment when international coordination in the field appears to be of growing importance, aiming at making data available as widely as possible to the astronomical community.

Acknowledgements

We would like to acknowledge the support provided during the editing of this book by ESA's European Space Research Institute (ESRIN), and by Observatoire astronomique de Strasbourg. The contribution of André Heck is also gratefully acknowledged.

Strasbourg and Frascati
February 1991

1

Data from the ROSAT Observatory

---

*H. U. Zimmermann & **A. W. Harris[1]
*Max-Planck-Institut für Extraterrestrische Physik
Karl–Schwarzschild Straße 1
D–8046 Garching bei München, Fed. Rep. of Germany
**Space Astronomy Group
Rutherford Appleton Laboratory
Chilton, Didcot, United Kingdom

1–1 Introduction

On June 1st, 1990, shortly before midnight, the ROSAT observatory was launched successfully into a circular orbit of 575 km altitude and 53 degrees inclination. The satellite carries two telescope systems on board: an X-ray telescope, operating in the energy range 0.1 to 2 keV, and a smaller XUV telescope sensitive in the range from about 20 to 200 eV. ROSAT —an abbreviation for Röntgen Satellite— is a German project with major contributions from the United States and the United Kingdom. Specifically the US (NASA/SAO) provided one of the 3 focal detectors of the X-ray telescope and launched the satellite, and the XUV telescope was built by a consortium of five British institutes, namely The Universities of Leicester and Birmingham, the Mullard Space Science Laboratory, Imperial College of Science Technology and Medicine and the Rutherford Appleton Laboratory.

On July 30, 1990, ROSAT started its main scientific task: to perform the first all sky survey with an imaging telescope both in the soft X-ray and the XUV regimes. After this survey, which will last 6 months, an extended observatory program will be performed (on a guest observer basis) with

---

[1]Present Address: WFC Quick-look Facility, MPE, D–8046 Garching bei München, Fed. Rep. of Germany.

pointings on individual targets. The wealth of information expected may be summarized as follows:

- detection of the order of 100,000 X-ray sources (presently less than 10,000 are known), comprising all object types from normal stars up to distant quasars,
- detection of 2,000 - 10,000 new EUV sources (before the start of the ROSAT survey only a handful of EUV sources were known).
- statistical studies of different object classes; because of the large numbers involved also classes of rare objects can be investigated,
- detailed studies of a large number of individual objects with regard to spectral and timing behaviour including observations at other wavelengths,
- mapping of the diffuse X-ray and EUV backgrounds with high sensitivity and spatial resolution.

1-2 The Instruments

The X-ray telescope (XRT) consists of a fourfold nested Wolter 1 type mirror with an entrance diameter of 80 centimeters and 3 focal detectors (Pfeffermann et al. 1986). Two of these are position sensitive proportional counters (PSPC) with a positional resolution of about 20 arcsec and a moderate energy resolution giving 4 to 5 bands within the 0.1 to 2 keV energy range. This detector has a field of view of 2 degrees and will be used during the all sky survey. The third detector is a High Resolution Imager (HRI) channel plate detector allowing resolution of a few arcsec without energy information. It is a rebuild of the HRI on-board the *Einstein* observatory.

The XUV telescope or Wide Field Camera (WFC) is a Wolter 1 telescope with 2 channel plate detectors in the focal plane (Barstow *et al.*, 1988). The positional resolution is about 1 arcmin. in the center of the 5 degree-wide field of view. Filters defining 4 wavelength bands are used to obtain spectral information.

1-3 Satellite Data Flow

The detectors in the focal planes of the X-ray and EUV mirrors record each photon arriving in the detector separately. The PSPC instrument, for example, delivers a 2-dimensional position of the photon event in detector coordinates, the pulse-height of the event as a measure of its energy and the arrival time. The data of each event are immediately written to the onboard tape recorder together with auxiliary information from different subsystems (housekeeping and attitude data). During the 5 to 6 contiguous contacts per day between the satellite and its ground station —the German Space Operation Center (GSOC) at Oberpfaffenhofen/Weilheim— the on-board recorded tape data are transmitted to the ground. About 20% of the data are immediately passed on via a dedicated line to the ROSAT Scientific Data Center (RSDC) for instrument health and data quality checks. The full set of telemetry data plus additional auxiliary information on orbit, attitude, commands and the mission program are delivered to the RSDC and, in the case of WFC data, to the WFC data centre in the U.K. a few days later.

Besides telemetry and command handling GSOC also implements the mission timeline and provides the primary attitude solution.

1-4 The ROSAT Scientific Data Centre

Located at the Max Planck Institut für Extraterrestrische Physik (MPE) at Garching, the RSDC is the centre for all interfaces between the observatory and the observers, the main data centres in the UK (RAL), the USA (GSFC) and at the German WFC data centre (AIT, Astronomisches Institut der Universität Tübingen).

The WFC Quick-look Facility (QLF), which forms a part of the RSDC, is responsible for monitoring the health and scientific data return of the WFC and compiling a first-cut EUV source catalogue. The Quick-look Facility, manned by WFC project personnel, also serves as the interface between the German and U.K. ground systems.

The tasks of the RSDC include (Zimmermann *et al.*, 1986):

- full responsibility for all XRT program activities, providing organisational, editorial and technical support during the *call for proposals*, the proposal selection and the mission timeline production phases;
- XRT instrument control and detailed checkout of all relevant XRT data in real and near real time;
- WFC instrument control and processing (by the QLF) of all WFC housekeeping data and 20% of WFC science data (about 10 Mb per day) for instrument health checking and monitoring of data return;
- standard data processing of all XRT data to achieve a uniform calibration and provide observers with the results of a standard image and source analysis;
- extended observer support, offering advice, documentation and visitor service, providing users with sophisticated analysis software, as well as working facilities at the RSDC and to retrieve data and results from the ROSAT Archives.

About 40 computers and workstations form the computational hardware basis for the over 30 persons who are presently providing this service to the scientific community.

1–5 Standard Data Processing

In the case of the XRT, the basic data sets for scientific data evaluation are produced through automated Standard Processing applied to all data from the XRT instrument. During that process the basic event information first undergoes various corrections and normalizations (*e.g.* detector linearization and pulse-height calibration). At the end of this step each event in the PSPC detector is represented by a 7 parameter set. In addition to the original x,y detector coordinates and the measured pulse-height and the arrival time, the corresponding x,y sky position (where the photon is most likely originating from) and a calibrated pulse-height (representing the most likely photon energy) are added. So called Photon Event Files containing this information form the basic input to all further analysis.

As the next step in the Standard Processing, images of the different sky regions are formed. Different techniques (sliding window and maximum likelihood methods) are used to search the images for point-like sources above a specified significance level. Probable values for the positions of detected sources, the sizes of the corresponding error boxes, the intensities in different energy bands and a source extension indicator are the output of the final maximum likelihood algorithm. For stronger sources a search for time variability is performed and spectra are derived and fitted with standard emission models. Finally, large optical catalogues are scanned for possible counterparts of the detected X-ray objects. Summaries of the results of the Standard Processing are produced and distributed —together with the standard data sets— to the relevant ROSAT observers by the ROSAT Archive service.

Processing of WFC data carried out at the Quick-look Facility and at the WFC project centres in the U.K. is similar to that performed on XRT data. Automatic pipeline procedures are used for both housekeeping and science data processing. After initial attitude and calibration processing, the WFC survey event data are added to a data reservoir held on disk which at any time contains data from the 360 degree band of sky being covered by the 5 degree-wide WFC scan path at that time. As the scan path precesses (at the rate of 1 degree/day), regions of the sky for which coverage is complete are 'closed off' and corresponding image data-sets (so-called 'small-maps') are extracted from the reservoir, searched for point sources and then archived to tape. The small-map scheme, which is similar to that used by the *Einstein* and *EXOSAT* projects, will also be used in the pointed phase. The pointed observation small-maps will be compatible with those from the survey. In addition to their basic small-map image data, guest-observers will be provided with standard data products of the kind described above for the XRT.

1–6 Extended Data Evaluation with EXSAS

It has been widely recognized in the past that archives containing raw or preprocessed data can only be effectively used if appropriate analysis software is commonly available. For ROSAT data the answer to that requirement is called Extended Scientific Analysis System (EXSAS). EXSAS has been developed by the RSDC for interactive analysis of X-ray data in general, with particular emphasis on the analysis of data from the ROSAT XRT and WFC instruments. This large collection of application software modules is embedded in the well known astronomical image processing system MIDAS, developed, maintained and distributed by the European Southern Observatory (ESO). EXSAS/MIDAS is highly portable and has been installed on both VMS and UNIX operating systems. In order to guarantee computer system independence regarding data distribution, all data output from the ROSAT Archives to the requesting observer are delivered as so called ROSAT Observation Data sets (ROD) in FITS format (a data transfer standard widely accepted by the astronomical community: see chapter 25). The ROD consists of more than a dozen tables and images containing all the basic and auxiliary data needed for further detailed analysis of ROSAT data: primary photon event data, information on instrument housekeeping and quality parameters, attitude and orbit files, calibration data and also selected results from the Standard Analysis performed at the RSDC.

EXSAS software may be requested from the RSDC by ROSAT observers having access to a MIDAS installation.

1–7 Archiving and Retrieval of ROSAT data at the RSDC

7.1 The ROSAT Archives

The ROSAT Archives at the RSDC consist of three major divisions:
- the ROSAT Data Archive contains all primary and secondary data products to be used for further processing,
- the ROSAT Result Archive consists mainly of selected results and data sets from the Standard Processing,

- an Astronomical Catalog and Skymap Archive holds major astronomical databases in use for the ROSAT source identification programs.

The amount of data to be archived lies in the range of 0.5 GBytes per day. In addition to the operational lifetime of the satellite, the final size of the archive will depend also on the rate at which reprocessing of data will be necessary. Assuming a lifetime of at least 3 years and 1.5 reprocessing steps (full and partial reprocessing) a planning figure for the final archive size of 2000 GBytes of storage capacity has been adopted. With this amount of data it is clear that only optical disk technology will enable the handling effort to be kept at a tolerable level. The RSDC has installed RV20 disk drives from DEC which can hold 2 GBytes of data on a single disk. The Archive Management System controls the offline archive on optical disks as well as a number of online magnetic disks (present capacity: 7 GBytes) as staging devices both for intermediate storage and for data sets that are needed online.

The underlying Database Management System (DBMS) used at the RSDC is INGRES. Sophisticated interfaces to both the requestor (to provide immediate status reports), and the operator (handling the disk drive loading/unloading), have been designed and (so far partially) implemented. Transaction handling, separate logging files and self-describing header information on the storage medium are the methods applied in order to facilitate reconstruction in the event of unexpected data loss due to system failure.

7.2 RSDC Archive Storage Considerations

There are clear differences between the different archive divisions in the way data are stored and retrieved.

The storage scheme of the ROSAT Data Archive directly reflects and supports the Standard Processing scheme described above. One part of the archive contains the original telemetry and auxiliary data as well as any secondary products that later on may be needed to reprocess data, starting either from the beginning or from major intermediate stages of the Standard Processing. Another part consists of all those data sets which are needed for users to proceed with their analysis. Some of the results of the Standard Processing are also stored here. This set of files exists in two flavours: first in the form as delivered by the Standard Processing, and secondly, slightly restructured for direct use within EXSAS, as FITS formatted ROSAT Observation Data sets. Further subdivision originates from differences while accessing pointed observation or survey data sets. The storage structure for the ROSAT Data Archive is the VMS Backup format. Whole directory contents are stored as one backup saveset containing the basic data sets together with all the auxiliary and supporting data files necessary for a specific kind of processing. This storage philosophy provides a compact and fast access to data. It is thus possible to retrieve with only one archive request all the files necessary for further processing/reprocessing by either the Standard Processing or the observer software system (EXSAS). For the ROSAT Data Archive therefore the DBMS holds, with a few exceptions, only the names, locations and configuration details of the directories containing the stored data sets.

In contrast, the storage method in the ROSAT Result Archive is based primarily on individual files. In addition, an appreciable part of the available information is directly stored in DBMS tables. The advantages of the relational DBMS can thus be fully used to search, access and present any combination of information and data items.

Access to the Astronomical Catalog and Skymap Database is based on a scheme that is strongly oriented on the requirements for fast retrieval of data from a specific region of the sky. The size of this archive (some tens of Gigabytes) prohibits holding it permanently on online disks. It is planned to keep about a 10% portion of the data loaded on magnetic disk at any time for online access and to swap it with other parts of the archive (different sky areas) according to requirements. Users of the Astronomical Catalog and Skymap archive are mainly the Standard Processing system and other software packages dealing with the requirements of the ROSAT source identification efforts.

Later on also the ROSAT catalogs and sky maps will be archived in this division.

7.3 Future Plans at the RSDC

Up to the end of 1990, complementing the present installation of the ROSAT Data Archive will be the primary task. In parallel, the design and implementation of the ROSAT Result Database will be started. The installation of the Astronomical Catalog Archive is already under way and should be finished before the end of 1990. With the growth of the archives, an extension of the presently planned retrieval procedures for users outside the RSDC has to be considered in 1991, taking into account first experiences obtained with the present installation.

As far as hardware is concerned the use of optical disks with higher capacity, drives with faster access times as well as the use of juke box systems for an highly automated archiving and retrieval environment will be considered as soon as relevant technology and expertise becomes available.

1–8 WFC Data

All WFC data are sent on Exabyte tapes by a daily courier service from GSOC to the U.K. Data Centre (UKDC) at the Rutherford Appleton Laboratory. The UKDC performs initial attitude and calibration processing and, during the survey, distributes pre-processed event, attitude and reduced housekeeping data files to the 5 WFC consortium sites participating in the WFC project. Each site, using common software, merges these data into its disk-held data reservoir and, as the survey progresses, produces source lists from 'closed-off' regions of that part of the sky for which it is responsible. Catalogue cross-checking is a further operation which is carried out automatically in the survey pipeline procedure at each site. After completion of the survey, the source lists from each consortium site will be merged to form a complete catalogue of EUV sources for the whole sky. The UKDC archives all the raw (50 Mb per day) and pre-processed data onto optical disk, while the medium used at most consortium sites for archiving the survey image data is Exabyte tape. The image data are archived as 2×2 degree small-map files in daily directories, each of which contains data for the region of sky which became closed-off on a particular day. The daily directories contain both raw and background- corrected maps, and the results of point-source searching and catalogue cross-checking.

During the pointed phase, the UKDC will be responsible for distributing data sets to guest-observers. In addition to the raw data, guest-observers will receive output from a standard analysis of their data. This will include an observation summary together with some basic scientific results which will act as a guide to further analysis. Software for more detailed analysis of the data will be available to guest-observers in the U.K. via the STARLINK astronomy network. In addition, a Guest Observer Centre, situated at the University of Leicester, will provide dedicated data analysis facilities and experienced staff to support guest-observer work and provide advice on instrument characteristics *etc*.

In the future, once data enter the public domain, it is likely that an archive of ROSAT data will be set up, probably based on optical disks. Small data sets may be accessible via the STARLINK network, while larger amounts of data could be requested on magnetic tape.

1–9 Archive Access and Data Rights

An important aspect of the XRT and WFC archive management systems is the access control of data from the pointed and the survey phases of the mission. Evaluation of XRT survey data can only be performed at the RSDC and in a collaborative approach with MPE scientists. After MPE has published results from the survey they will also be accessible from the ROSAT Archives in form of catalogs and sky maps.

For the pointed observation phases, proposals can be sent to one of the data centers in Germany, the UK or the USA. Data from approved observations will then be distributed automatically to the corresponding Principal Investigator as soon as they become available. These data belong exclusively to the proposal originator for one year and will enter the public domain after that.

Due to these rules, requests for data and/or results to the ROSAT Archives cannot be accepted before the above mentioned property rights on the data are expired. With the present planning the first pointed observation phase of ROSAT will begin in February 1991. Therefore in the course of 1992 the ROSAT Archive Services will begin to distribute on request ROSAT data and results also to the general astronomical community.

The ROSAT Scientific Data Center is planning to provide ROSAT Archive Services also on a long term basis in the post mission era. In addition data archives will be built up both in the US (GSFC/NSSDC) for ROSAT X-ray data, and in the UK (RAL) and at the German WFC data center (AIT) for ROSAT XUV data.

References

[1] Barstow M. & Willingale R. (1988), *JBIS*, **41**, 345.
[2] Pfeffermann E., Briel U. G., Hippmann H., Kettenring G., Metzner G., Predehl P., Reger G., Stephan K.-H., Zombeck M. V., Chappell J., Murray S. S. (1986), *SPIE*, **733**, Soft X-Ray Optics and Technology, 519.
[3] Zimmermann H. U., Gruber R., Hasinger G., Paul J., Schmitt J., Voges W. (1986), Data Analysis in Astronomy II, 155.

2

The EXOSAT Database System

N.E. White & P. Giommi
EXOSAT Observatory
Astrophysics Division
Space Science Department of ESA
ESTEC

2–1 Introduction

The European Space Agency's X-ray observatory EXOSAT was operational from May 1983 to April 1986 and made 1,780 detailed observations of a wide variety of objects including active galactic nuclei, stellar coronae, cataclysmic variables, white dwarfs, X-ray binaries, clusters of galaxies and supernova remnants. The payload compliment consisted of a large area proportional array, two imaging telescopes (each with a transmission grating spectrometer), and a gas scintillation proportional counter. A unique and highly successful aspect of the mission was the eccentric 90 hour orbit of EXOSAT, which gave for the first time the capability to make uninterrupted observations of the time variability of X-ray sources for several days at a time. The proprietary data rights of an observer were restricted to the first year after data receipt, and by the end of 1987 the entire EXOSAT data archive was open to the astronomical community.

The *EXOSAT database system* utilises the recent expansion of computer networks and advances in database management techniques to provide on-line access to EXOSAT data and results. World-wide, real-time access is provided by the Space Physics Analysis Network (SPAN), the Internet (TCP/IP) and public X.25 connections. Ultimately the EXOSAT database will become a node within the European Space Information System (ESIS).

2–2 A Data Archive Infra–structure

The EXOSAT mission produced a total data volume of 150 Gigabytes (6,000 magnetic tapes). As the data became freely available in 1986-1987 there was a surge in archive requests resulting in the dispatch of 600 final observation tapes in 1987 (or 10% of the entire archive). Since then this has settled down to a steady rate of 350 tapes per year in 1988 and 1989.

The reduction of these raw data tapes into data products (images, light-curves and spectra) is complex because of the sophisticated nature of the instruments and on-board data compression techniques. Extensive analysis software, and expertise in the nuances of the instrumentation is required to obtain high quality data products and results. While several groups in Europe have made such an effort, mostly those associated with the EXOSAT instrumentation, many others do not have the required resources.

The complex nature of the EXOSAT data analysis acts as a barrier to archival research. Past experience with complex astrophysics missions like EXOSAT had shown that after only a few years, the data becomes inaccessible as the expert personnel move on to new projects. In addition, it is well known that only a fraction of the data can be fully exploited while the spacecraft is operational, and that the scientific return of the mission increases substantially when the data is made available to a wide scientific community. A high priority of the EXOSAT post-operational phase was to establish an archival infra-structure to ensure long term easy accessibility to the EXOSAT data archive.

Many of the required data processing tasks are repetitive and amenable to automation. A number of standard data products (light-curves, images and spectra) and results (count rates, source positions, *etc.*) were defined for the EXOSAT mission and then generated by running the EXOSAT analysis software in an automatic mode. The EXOSAT database system is designed to allow non-experts with only a rudimentary knowledge of the EXOSAT instruments, *e.g.* a theoretician, to immediately gain access to these data products and results.

The available data products are limited by the available magnetic disk space, a volume of 9 Gigabytes. They will mostly consist of low time resolution spectra, light-curves and images, plus high resolution light-curves from a selection of objects. It is obviously impossible to anticipate every data product that will be required and in some cases, an astronomer must still request and analyse the raw data tapes. This is expected to be in approximately 20% of the cases. But even then the EXOSAT database is invaluable in establishing the available observations and overall source characteristics. Thus the astronomer can minimize the analysis of the raw data, and hence the time spent at institutions with a full EXOSAT analysis system.

Data products and catalogs of results from other related observatories are also made available through the EXOSAT database. These include data from NASA's Einstein observatory, IUE/ULDA spectra of sources observed by EXOSAT, and various astronomical catalogs *e.g.* the Hubble Space Telescope Guide Star Catalog. These related data are essential for making correlated studies, source identification, and long term variability studies.

2-3 The User Interface

The user interface to the EXOSAT database system is a command driven environment optimised for astronomical queries that allows astronomers, both experts and novices, to *browse* through the available data, and select those required. The user interface was developed with the active participation of the astronomical community. Two "database workshops" were held at ESTEC in December 1988 and February 1990 to demonstrate the system and obtain feedback to improve it. Forty astronomers attended each workshop. Via this process an astronomical query language optimized for both browsing through the database, and for remote access has been developed.

Graphics is an important element that allows the display of plots and images directly over the networks. Figure 2-1 shows a typical spectrum from the EXOSAT database system.

Figure 2-2a,b show a plot in galactic coordinates of all the sources detected by the CMA and LE instruments. This on-line graphics capability allows the astronomer to directly visualise the results and data products. All popular graphics devices are supported. In addition postscript files can be produced and electronically mailed to the user, to give publication quality output. The astronomer can make further analysis on the data products either by copying them over the networks, or by running the analysis programs available on the EXOSAT system.

A database management system (DBMS), is used to locate the results and data products. The DBMS has been written by members of the EXOSAT observatory team and is optimized for astronomical queries. Here the emphasis is orientated to correlation analysis between large tables of parameters that are rarely changed (*e.g.* a catalog of stars). The EXOSAT database operating system has been specifically written in a mission (and wavelength) independent fashion and can be applied to other astrophysics domains.

The EXOSAT database provides a direct interface between the data products and the analysis software. This allows remote users to analyse

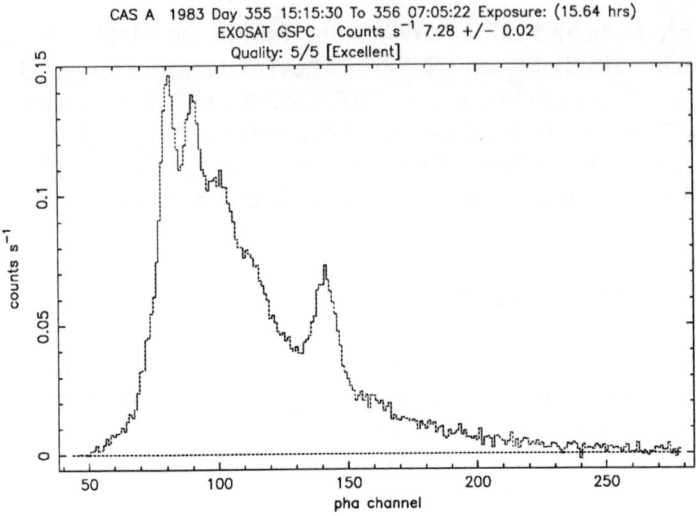

Figure 2-1: EXOSAT Spectrum of Cas A

thousands of data products, selected using the DBMS. These products can be further selected based on the results of this analysis. In this way the EXOSAT database system provides a powerful tool for the systematic analysis and interpretation of large data volumes.

The EXOSAT database system was made available for remote use in April 1989. In the first year of use a total of 182 remote users had logged-in to the system with a total of 3,261 sessions. Figure 2–3 shows the usage per week. Currently there are typically 5 remote users per day, with a total of 50-100 sessions per week.

2-4 The Future

The EXOSAT database system has applied database management technology to data analysis in the general astrophysics context and in doing so has surpassed the original design goal of simply managing the EXOSAT data products and results. It is a new generation of data handling and analysis system optimized for large data volumes and network access. Externally defined data structures specify the underlying binary data so that virtually any data format can be incorporated, without any changes to the underlying software. This will allow the analysis of data from many different missions within a single analysis environment. Indexing of the raw data

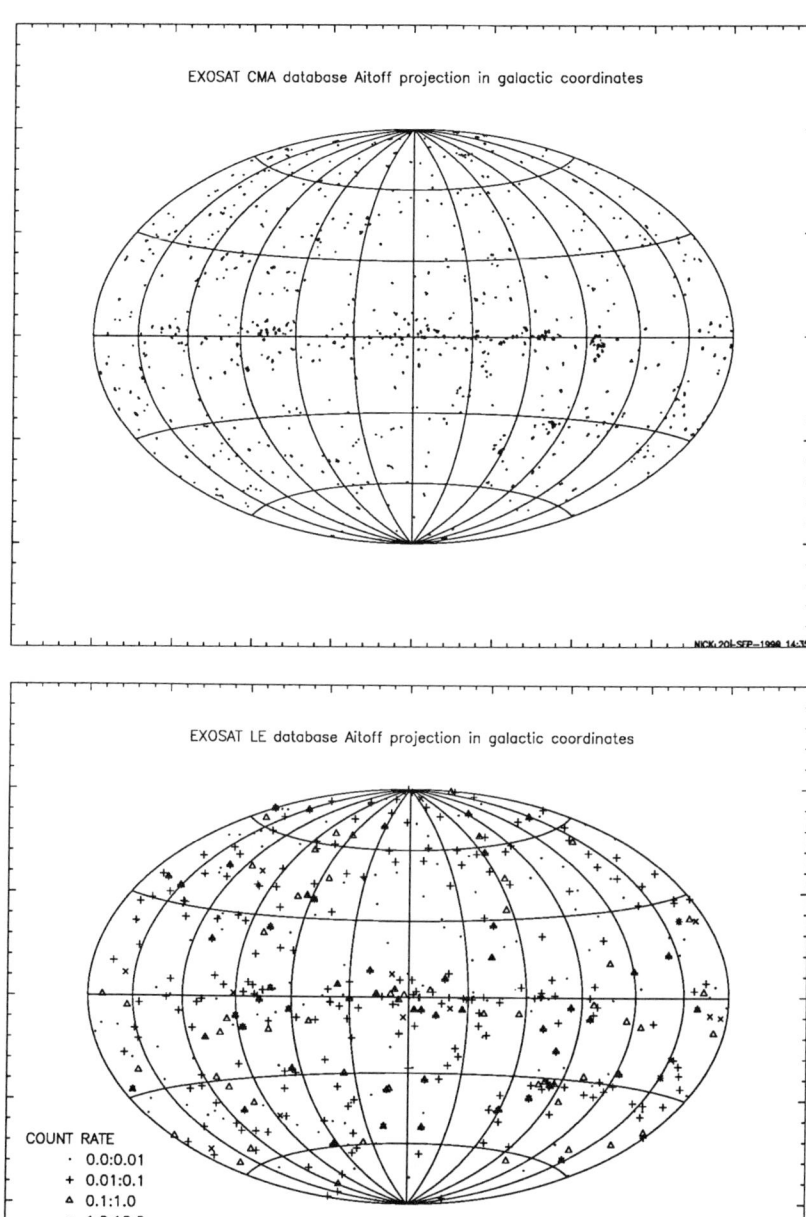

Figure 2–2: X-ray sources observed by EXOSAT CMA (a) and LE (b) instuments.

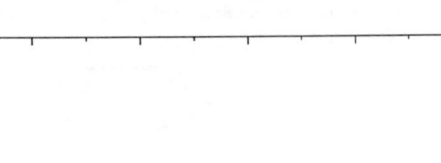
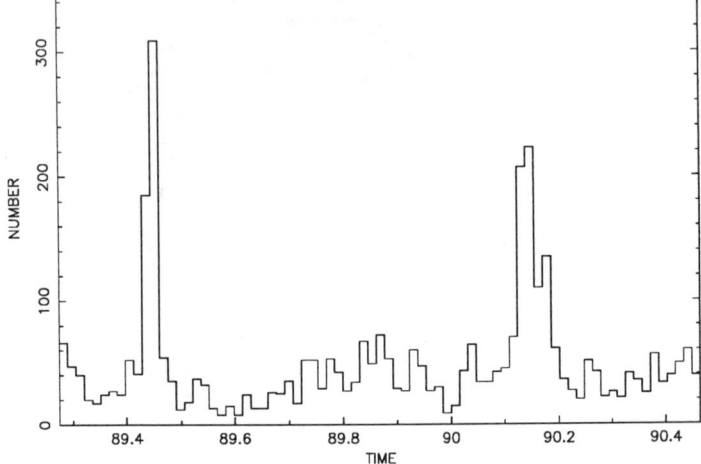

Figure 2-3: Usage of the EXOSAT database in number of remote connections per week.

on any constituent parameter will give rapid data access.

The EXOSAT database system will be released in the latter half of 1990 as a package. It will be installed at sites around Europe and the rest of the world, as a complete data archive/data analysis system. It is expected that these sites will install a subset of the EXOSAT data products that are of particular interest to that institute, and make them available for local use. The X-ray astronomy group at Leicester University, and the NSSDC at the Goddard Space Flight Center are testing the current system, with a view to its use in their own archiving efforts.

2-5 Conclusions

It can be anticipated that in near future astronomers will access vast amounts of astronomical data, both new and archival, from terminals/workstations located thousands of kilometers from where the data are collected and stored. Such a capability will only be of use if the data are organised in a manner that aids the enquiring astronomer to rapidly search, filter and display catalogs, data products and raw data based on any combination of parameters. The EXOSAT database system provides the required infrastructure.

3

IRAS

Helen J. Walker
Rutherford Appleton Laboratory,
Chilton, Didcot, Oxon, OX11 0QX,
United Kingdom

3-1 Introduction

The InfraRed Astronomical Satellite (IRAS) was a joint American, Dutch and British project. The satellite was launched in January 1983 to survey the whole sky in the infrared, specifically at wavebands centred around 12μm, 25μm, 60μm, and 100μm. The satellite was cooled using liquid helium, and the mission ended in November 1983 when the helium ran out and the telescope started to warm up. In addition to the 62 main survey detectors, the satellite contained a Chopped Photometric Channel (CPC) sensitive at 50μm and 100μm, and a slitless Low Resolution Spectrometer (LRS) sensitive between 7.7μm and 23μm. About 60% of the mission time was used for the all-sky survey, leaving around 40% of the time for engineering tests, calibration observations and pointed observations of interesting objects. It was planned to release a catalogue of point sources, a catalogue of sources which were slightly extended, a catalogue of spectra, and small maps (in FITS format) of the pointed observations. In fact, many other data products are available, and due to new image processing techniques, high quality images using the IRAS survey data have been produced.

1.1 The IRAS satellite

The IRAS satellite consisted of two main parts, the spacecraft (supplied by the Netherlands through the Industrial Consortium IRAS, involving Fokker, Signaal and the National Aerospace Laboratory) and the telescope (supplied by the U.S.A. through the Jet Propulsion Laboratory (JPL), NASA Ames Research Center, Ball Aerospace Systems Division, Rockwell International and Perkin Elmer). The satellite was 3.60m high, and 3.24m wide with the solar panels deployed. At launch it weighted about 1000 kg. The Ritchey-Chrétien telescope had a 60cm primary mirror, and a 63.6 arcmin unvignetted field of view. The focal plane contained the 62 survey detectors, 8 visible star sensors, the CPC and the LRS. Each of the four wavebands had its detectors arranged in two modules, one on each side of the focal plane, so that an initial condition for deciding whether a source detected by IRAS was real, was that it should be seen by both modules (see the IRAS Explanatory Supplement, 1988, for the layout of the detectors in the focal plane). The survey detector array covered an area with diameter about 30 arcmin on the sky. The detectors were kept at a temperature of 2.6K, some of the electronics were at around 70K to 80K, and the temperature at the outer shell of the telescope was around 195K.

1.2 The survey

The satellite orbitted the earth 14 times a day, at an altitude of 900km. It surveyed along the day/night line on earth (the terminator). It was not allowed to point closer than 60° towards the sun, or more than 120° away from the sun, and was not allowed to look at the Earth, the Moon, or Jupiter. The satellite scanned the sky at 3.85 arcmin/second, and the survey scans were laid down so that each scan half-overlapped with the previous scan. This gave a second level of confirmation on the reality of the sources. Every two weeks the area of the sky being scanned was moved, and this again half-overlapped with the area previously scanned, giving a final level of confirmation for the source. In this fashion the whole sky, apart from a strip of about 5°, was surveyed in 6 months. When this initial survey was finished, it was decided to survey the sky with a single extra coverage (*i.e.* not overlapping after two weeks), and 76% of the sky was surveyed in this way before the helium ran out.

The data from the detectors were stored on tape recorders on the satellite and down-linked twice a day to the control centre at Rutherford Appleton Laboratory (Chilton, U.K.). Comprehensive checking was carried out to ensure that the satellite was working satisfactorily; if it was not, corrective measures were speedily developed and uplinked. The data were then copied to the Jet Propulsion Laboratory (Pasadena, U.S.A.) for eventual calibration, source extraction, and catalogue production. In order to check on the scientific quality of the data, it was necessary to perform a rapid scientific analysis of the sources seen by IRAS. Among the software tools developed for source extraction and analysis at RAL, was the capability to compare sources extracted from the IRAS data stream with known sources, from catalogues at other wavelengths. For this the Catalogue Access and Reporting software was developed, which later became SCAR (see the chapter on Starlink, page 165, for a detailed description). The same concept of comparing sources found in the IRAS data with known sources was used at JPL, and the final IRAS catalogues cross-reference the information. The catalogues used at RAL were supplied by the Centre de Données astronomiques de Strasbourg (CDS), and those used at JPL were provided by the NSSDC (see chapters 9 and 16). About 40 catalogues were used for the comparison, and no attempt was made to limit the association of an IRAS source with any catalogue, except that IRAS sources seen only at 100μm were assumed not to be stars.

1.3 Source extraction

In order to understand the origin of the data products, it is necessary to explain further how the sources were extracted from the data streams. A zero-sum point-source template was moved across the data stream from each detector, and the position and intensity of any "source" stored. The thresholds for registering a detection were deliberately set to low values, so that the completeness of the catalogue was maximised. However, many spurious detections naturally resulted from this, and since only genuine astronomical sources were to be published (*i.e.* excluding even asteroids and comets), the source detections were rigorously checked to ensure they were reliable (and the spurious sources were not stored). The several sightings of the source were used to check its reality. If the source was seen, at a particular wavelength, by both sides of the focal plane (with the correct time delay) it was called a *seconds-confirmed* detection. At this point the detections for all four bands (12μm, 25μm, 60μm, 100μm) were compared and if the positions fulfilled various tests, the detections were combined to give a seconds-confirmed band-merged source, with a refined position. On the next orbit (100 minutes later) another pair of seconds-confirmed sightings would be recorded. Sources from these two sightings (hours apart) could be combined, if positional and flux criteria were satisfied, to give an *hours-confirmed* source. Finally a third level of confirmation (*weeks-confirmation*) was required before the source was allowed into the pool of sources from which the catalogue would be produced, and this was achieved using the data collected two weeks later, when the survey area had been moved. At this point all the individual detections of the source were retained, and the source was deposited in the [IRAS] Working Survey DataBase (WSDB). Originally there were no plans to release the WSDB, but a subset of it was made available with Version 1.0 of the IRAS Point Source catalogue, and the whole WSDB was released by JPL to the astronomical community in November 1986 (Lonsdale, 1989).

3-2 Data products

Infrared astronomers were very aware of the enormous increase in information provided by IRAS. Figure 3-1 shows a comparison of the situation at 11μm before and after IRAS. The data in the top panel came from the compilation of Gezari *et al.* (1984), and show all the observations made prior to the launch of IRAS at 11μm, the longest infrared wavelength (easily) accessible from the ground, around 2310 sources in all. The lower panel shows point sources found at 12μm by IRAS; around 154,000 sources. These plots were produced using SCAR on Starlink (see chapter 18). A relational database, such as SCAR, made the exploitation of the new information a lot easier, in that subsets of sources could be quickly extracted from the catalogues using a wide variety of criteria.

The other overwhelming aspect of the data was the quantity of information available about each source. There were coordinates and four flux densities given for each source (although some might be upper limits); also usually available were associations with objects in other catalogues, spectra, background estimates, flags to warn about nearby sources and other processing problems. Table 3-1 (see, for more information, the IRAS Explanatory Supplement, 1988), shows the amount and style of information in the IRAS Point Source catalogue, and this does not specify explicitly *all* the information available on magnetic tape for each source!

The IRAS catalogues and images are available on magnetic tape, through the two major catalogue distribution systems, NSSDC for the U.S.A. and CDS for Europe. Software to read the tapes is not the responsibility of these centres. The IRAS Point Source catalogue is also available as a printed book, as are several other catalogues. The normal procedure for releasing a new product is that IPAC (Infrared Processing and Analysis Center, CalTech) sends magnetic tapes to NSSDC, and certain other centres (such as RAL and the University of Leiden in the Netherlands), and releases a hardcopy *Explanatory Supplement* describing the catalogue and the processing done to achieve it.

2.1 The IRAS catalogues

Table 3-2 shows the currently available IRAS catalogues (as of late 1990), the information is taken from the IPAC Newsletter (Lonsdale, 1989). A unique aspect of IRAS is that it was agreed to reference the primary catalogues as coming from IRAS rather than from people, hence some catalogues in the references are listed under their title instead of using the names of their editors.

Parameter	Description
Name	IRAS name of source
Position	RA and Dec (1950.0)
Position uncertainties	Not in RA and Dec originally, but converted
Fluxes	IRAS flux densities or upper limits at 12μm, 25μm, 60μm, 100μm
Flux qualities	Whether upper limit or not, for each flux
HCONS	Number of times the source was seen
LRS spectra	Number of LRS spectra used in Atlas mean
LRS flag	Classification of LRS spectrum
Flux uncertainties	One for each band
SNR flag	SNR in each band ($\times$ 10)
Correlation coefficient	How well it matched the point source template
Variability	How well the several sightings matched
Discrepancy flag	Were any fluxes discrepant?
Confusion flag	Was there a lot of extended emission nearby?
High density flag	Were there a lot of sources nearby?
SES flag	Did the source appear extended?
Associations flag	How many other catalogues have a source there?
Type flag	What sort of catalogues associated?
Catalogue number	Associations catalogue identifier (for each)
Catalogue name	Name of source in associations catalogue (for each)
Fields	Up to three pieces of information from the catalogue

Table 3–1: Contents of IRAS Point Source catalogue

Product	Version	Release Date	Book	Tape
IRAS Point Source catalogue	2.0	Nov 1986	*	*
Small Scale Structure catalogue	1.0	Dec 1985		*
IRAS LRS Atlas	1.0	Nov 1984	*	*
Serendipitous Survey catalogue	1.0	1986		*
Faint Source Survey $|b| > 10°$	2.0	Sep 1990		*
IRAS Asteroid and Comet Survey	1.0	Oct 1986	*	*
Catalogued Galaxies and Quasars	2.0	Mar 1989	*	*

Table 3–2: The IRAS catalogues

The IRAS Point Source catalogue (Version 2, 1988) contains 245,839 sources, of which almost half are thought to be stars and around 50,000 sources are thought to be galaxies. This catalogue is, however, only one of several catalogues released by the IRAS project. In addition to point sources, a catalogue of sources up to 8 arcmin in diameter was released (containing 16,740 sources), called the IRAS Small Scale Structure catalogue (1988). A subset (Catalogued Galaxies in the IRAS Survey, Version 2, 1989) of the IRAS Point Source catalogue was released containing those sources which were associated with galaxies and quasars, containing 11,444 sources, and in this catalogue there were not only IRAS point source data for the sources but also information from the Small Scale Structure catalogue.

The WSDB was used to extract spectra from the Low Resolution Spectrometer (LRS). Since the spectrograph was slitless, there were many problems with overlapping spectra and baseline subtraction. Consequently the criteria for inclusion in the final released catalogue were very strict. The Atlas of Low Resolution IRAS Spectra (1986) contains the averaged spectra of 5425 sources. The associations to other catalogues as well as priorities used for selecting the astronomically most useful name for the source from its associations are given in the LRS Atlas. Some estimate of the quality of the spectra is available, and the spectra were classified using a simple classification scheme, set up (of necessity) before IRAS was launched. This meant that any new information (discovered through the spectra) could not be used to revise the classification.

It was stated previously in this chapter that objects such as asteroids and comets would not be specifically included in the IRAS databases forming the catalogues. Some solar system objects could move so fast that they would fail to achieve seconds-confirmation. During the IRAS mission, special software was used at RAL to hunt for new comets and asteroids, so that they could be quickly observed using special pointed observations. IRAS found 6 new comets during its 10 month mission, and detected a large dust tail to the Tempel 2 comet. A new asteroid (1983TB) was also discovered by IRAS, in the same orbit as the Geminid meteor shower. For the IRAS Asteroid and Comet catalogue (1986) the positions of 1811 known asteroids and 25 comets were used to retrieve IRAS data for them. There are plans for a second version of the catalogue including asteroids and comets whose orbits are unknown, but that search through the database of unconfirmed IRAS detections is a subtle one. The criteria for confirming the reality of any potential source are very difficult to define, since the confirming positions are unknown in advance.

2.2 The IRAS images

Table 3–3 shows the officially released images from the IRAS data (see the IPAC newsletter, Lonsdale, 1989, for further details). The data from the 3 sets of hours-confirmed (HCON) scans were kept separate when the images were produced, since the solar illumination of the zodiacal dust was different at each time and, after six months, because the dust was so close, it showed parallax. All the images were produced in FITS format, except for the Zodiacal History File.

Product	Version	Release date	Pixel size
Skyflux plates			2'
HCON1	1.0	Aug 1985	
HCON2	1.0	June 1986	
HCON3	2.0	May 1986	
Galactic plane images			2'
HCON1	1.0	Oct 1985	
HCON2	1.0	June 1986	
HCON3	2.0	July 1986	
Faint Source Survey plates $|b| > 50°$	1.0	Dec 1988	0.25'/0.50'
Zodiacal History File	3.1	June 1989	30'
All-sky maps	1.0	Dec 1984	30'
Pointed observations	1.0	Oct 1985	various
CPC Data	2.0	Aug 1989	0.33'

Table 3–3: IRAS image products

The main IRAS image products in use are the pointed observations and the Skyflux plates. IPAC released the set of images of the whole sky arranged in 16.5° × 16.5° plates, at each of the four wavelengths. The concept of adding the individual detector scans together to make an image was used in the Netherlands during the mission as an additional tool to check on the quality of the data, and to find interesting extended infrared sources to follow-up with pointed observations. The IRAS calibration and the zodiacal light (from the warm dust in our own solar system) caused problems in the original Skyflux maps. These have now been resolved, and the second release (Super-Skyflux) will have a model of the zodiacal light subtracted from the IRAS data, improved flux calibration, and a slightly smaller pixel size (1.5').

The pointed observations have also been released as small maps, in FITS format. Each observation was performed at least twice, to give confirmation of the reality of features seen in the map. For the pointed observations done with the survey array, the maps were co-added and the point sources in the map extracted to provide the IRAS Serendipitous Survey catalogue (Kleinmann et al., 1986) containing 43,866 sources. The positions of the sources are less accurate, due to the nature of the data taking, but the limiting sensitivity is typically improved by a factor of four. Some pointed observations were made with the CPC instead of the survey array, since that had a smaller, circular beam.

2.3 Other IRAS data in use

The Calibrated Raw Detector Data (CRDD) are available in various places: IPAC, several Dutch institutes and in the UK, Queen Mary and Westfield College (London) and IRAS Post Mission Analysis Facility at RAL (IPMAF) . These data can be used, if the need arises, to look for weak sources, *etc..* IPAC have recently recalibrated the raw data, in preparation to the Faint Source Survey.

Rice et al. (1988) have released a catalogue of IRAS data on 85 large galaxies, with diameters greater than 8 arcmin in Blue light. This gathered together data from several IRAS catalogues, and measurements of flux densities from the CRDD.

The LRS Atlas (as mentioned in the previous section) had very strict selection criteria for the 5425 mean spectra to be included. The dataset of individual spectra (up to 170,000 spectra of potentially around 40,000 sources), on the other hand, could include sources that were not even in the IRAS Point Source catalogue, because they failed the final strict position and flux tests. RAL obtained a copy of this database from the Space Research Institute in Groningen (Netherlands), who had received their copy from the University of Leiden, which had responsiblity for the LRS database during the IRAS mission. RAL made the database available by putting it onto the Starlink database machine and users can interrogate it remotely with queries based on the astronomical coordinates of the source. Since it contains the individual spectra of the subset of sources which appeared in the IRAS LRS Atlas, checks can be made on the reality of "odd" features (*e.g.* de Muizon *et al.*, 1988). Volk and Cohen (1989) have used this database to extract (essentially by hand) the subset of bright IRAS sources (with flux density at $12\mu m > 40$ Jy) which had LRS spectra and yet were not in the LRS Atlas due to processing problems.

Cheeseman *et al.* (1989) using artificial intelligence techniques reclassified the IRAS LRS Atlas. For the spectra containing the $10\mu m$ silicate dust feature (around 60% of the spectra) a finer classification than before was possible, based on the shape, strength, and precise wavelength of the feature, and whether there were other weaker features in the spectrum or not. For classes of spectra with few representatives, the classification had

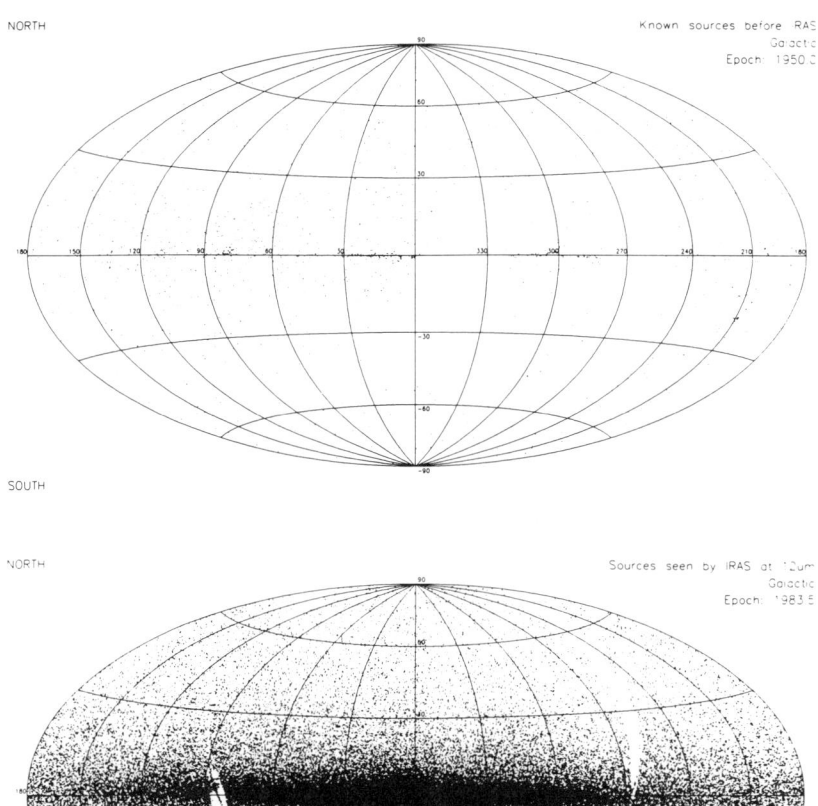

Figure 3–1: Known sources before and after IRAS

a comparable quality to the original classification.

3-3 Data access

It was expected that there would be interest in IRAS data long after the end of the mission, so that distribution centres for the data and software would be needed. In the USA, IPAC was set up, with priority at first resting on producing the official data products on time (as per an international Memorandum Of Understanding). US astronomers wishing to access the data apply to IPAC, as though to a ground-based telescope, and the applications are assessed by peer review. Astronomers usually go to IPAC to obtain their data and analyse them, with the assistance of IPAC personel. IPAC will retrieve data and mail them to astronomers, after their initial visit. In the UK, the mission control centre at RAL formed the nucleus of the UK distribution centre, creating IPMAF after the end of mission. The role of RAL had originally expected to cease when the helium ran out of IRAS, and the main-frame computer was to be scrapped. After consultation with Starlink, IRAS software, developed at RAL during the mission for scientific analysis, and data were moved to the Starlink VAX/VMS system and made available to UK astronomers. Documentation for the software was provided, so that users could easily access the data whenever they needed. The chapter on Starlink (page 165) explains the access in more detail. In the Netherlands the situation is again different. Groups in three universities were heavily involved in the IRAS mission, and they have developed their own software to handle the IRAS data which interest them. Dutch astronomers wishing to use the data, contact the most appropriate institute for their purposes, *e.g.* University of Leiden for spectra.

3-4 Future developments

The Space Research Institute in Groningen is developing an exportable package, called GEISHA, to analyse IRAS data using the GIPSY data structure. IPAC are still developing new data products. As has been mentioned previously, improved data processing and re-calibration have meant that the quality of the almost-raw data (Calibrated Raw Detector Data) has been significantly enhanced. These images will themselves be most useful, as will their derivatives, in particular the Super-Skyflux plates. The next generation of infrared satellites, ISO and SIRTF, which are observatories rather than survey instruments, will build on the IRAS database. This means that astronomers preparing to use them will be pushing the IRAS data to their limits (and probably beyond) of sensitivity and positional accuracy.

In the UK, the IRAS Starlink Applications programmer has applied Maximum Entropy techniques (Berry, 1990 and Gull & Skilling, 1984) to improve the resolution of IRAS data, and this software is available to all UK astronomers through Starlink. Starlink programmers (as mentioned in chapter 18) are moving software to the still-developing ADAM environment and are using HDS (Hierarchical Data Structure) files. The UK community has decided that the IPMAF software should move into this environment, so that a lot of re-coding will be necessary. This is an appropriate time to consider the needs of astronomers preparing to use ISO, and the software will be upgraded in the light of those requirements.

Acknowledgements

The InfraRed Astronomical Satellite was developed and operated by the Netherlands Agency for Aerospace Programmes (NIVR), the U.S. National Aeronautics and Space Administration (NASA), and the U.K. Science and Engineering Research Council (SERC). Thanks are due to the many people, who wrote software for IRAS before, during, and after its mission, to enable so many astronomers to enjoy the data.

References

[1] Atlas of Low Resolution IRAS Spectra (1986), IRAS Science Team, prepared by F. M. Olnon and E. Raimond, Astron. Astrophys. Suppl., 65, 607.
[2] Berry D.S. (1990), IRAS (UK) Newsletter No. 2, p8. Edited by H.J. Walker.
[3] Cataloged Galaxies and Quasars Observed in the IRAS Survey, Version 2 (1989), Prepared by C.J. Lonsdale, G. Helou, J. Good, W. Rice, (Pasadena, JPL).
[4] Cheeseman P., Stutz J., Self M., Taylor W., Goebel J., Volk K. and Walker H. (1989), Automatic Classification of Spectra from the Infrared Astronomical Satellite (IRAS): NASA Reference Publ. 1217 (Washington DC: GPO).
[5] Gezari D.Y., Schmitz M., Mead J.M. (1984), Catalog of Infrared Observations: NASA Reference Publication 1118 (Washington DC: GPO).
[6] Gull S.F. & Skilling J. (1984), IEE Proc. F. 131, No. 6, p. 646.
[7] IRAS Asteroid and Comet Survey (1986), Edited by D.L. Matson, (Pasadena, JPL).
[8] IRAS Catalogs and Atlases: Explanatory Supplement (1988), Joint IRAS Science Working Group, edited by C.A. Beichman, G. Neugebauer, H.J. Habing, P.E. Clegg and T.J. Chester (Washington DC: GPO).
[9] IRAS Point Source Catalog, Version 2. Joint IRAS Science Working Group (Washington DC: GPO).
[10] IRAS Small Scale Structure Catalog (1988), Prepared by G. Helou and D. Walker (Washington DC: GPO).

[11] Kleinman S.G., Cutri R.M., Young E.T., Low F.J., Gillett, F.C. (1986), Explanatory Supplement to the IRAS Serendipitous Survey Catalog (Pasadena, JPL).
[12] Lonsdale C.J. (Editor, 1989), IPAC Newsletter, Vol. 5, No. 1.
[13] Moshir M., et al. (1989), Explanatory Supplement to the IRAS Faint Source Survey Catalog (Pasadena, JPL).
[14] Muizon M. de, Cox P., Lequeux J. (1988), Astron. Astrophys., **203**, 207.
[15] Rice W., Lonsdale C.J., Soifer B.T., Neugebauer G., Koplan E.L. (1988), Astrophys. J. Suppl., **68**, 91.
[16] Volk K. & Cohen M. (1989), Astron. J., **98**, 931.

4

Databases from Cosmic Background Explorer (COBE)

[1]Richard A. White & [2]John C. Mather
[1]*Space Data and Computing Division*
NASA/Goddard Space Flight Center
Code 936
Greenbelt, MD 20771

[2]*Laboratory for Astronomy and Solar Physics*
NASA/Goddard Space Flight Center
Code 685
Greenbelt, MD 20771

4–1 The COBE Mission

The Cosmic Background Explorer (COBE)[1] was launched in November 1989. Its scientific objectives are to search for spatial anisotropies and spectral distortions in the 2.7 K cosmic microwave background (CMB) radiation, to detect the diffuse infrared background radiation from the first objects to form after the Big Bang, and to study all other sources of diffuse radiation from 1 micron to 1 centimeter. These other sources include interplanetary and interstellar dust, hot electrons in the Galaxy, faint stars in the Galaxy, and possibly IR galaxies and hot gas in galaxy clusters. To map these primeval and local sources, the three scientific instruments on COBE scan the sky repeatedly, building up signal-to-noise statistics until the data are limited only by the astrophysical environment. The three instruments are the DIRBE (Diffuse Infrared Background Experiment) covering 1 to 300 micron with a 10 band filter photometer and a 0.7° beamwidth; the FIRAS (Far Infrared Absolute Spectrophotometer) covering 100 microns to 1 cm with an absolutely calibrated polarizing Michelson interferometer with a 5

---
[1]COBE is a NASA space mission in the Astrophysics Division of the Office of Space Science. The COBE project is managed by the Goddard Space Flight Center.

percent spectral resolution and a 7° beamwidth; and the DMR (Differential Microwave Radiometers) covering 31.5, 53, and 90 GHz with 7° beamwidth. COBE is described further elsewhere (see Gulkis et al., 1990).

All three instruments have already made spectacular advances in the quality and quantity of the data available (see Mather et al., 1990), and are expected to lead to the discovery of new phenomena in the early universe. The principal scientific goals of each instrument are to make definitive measurements of the spectrum of the cosmic microwave background (FIRAS), to measure its large angular scale intensity distribution (DMR), and to obtain absolute measurements of diffuse infrared emission from sources in the early universe and local astrophysical sources (DIRBE). The data will be clearly useful for many other scientific studies as well.

COBE is in a near polar, sun-synchronous 900 km orbit which maintains the orbit plane on average at the day-night line of the Earth. The spin axis is kept pointed at a nominal 94° angle from the Sun, and as far from the Earth as is consistent with the 94° requirement. The FIRAS points along the spin axis. The DIRBE and the DMR are pointed 30° from the spin axis, and the spacecraft spins at 0.8 rpm. Thus DIRBE and DMR sample a 60° wide band on the sky with each orbit. For DIRBE, this allows the zodiacal light to be sampled at elongation angles from 64° to 124° at all points along the Earth's orbit. At the time this article is written (June 1990) the COBE has just completed one complete coverage of the sky. A second coverage is underway and it is hoped it will be completed before the liquid helium cryogen is depleted. At that time the FIRAS will completely cease to function but several of the DIRBE channels may still be able to provide data suitable for studies of the interplanetary dust. The DMR, which is not helium cooled, is expected to continue to operate for at least one additional year.

COBE was designed, built, integrated, and tested at the NASA Goddard Space Flight Center (GSFC) with scientific guidance from the Science Working Group (SWG). COBE operations in orbit and the final data analysis also take place at GSFC. The COBE SWG is responsible for the definition, integrity, and delivery of the public data products. The working group members and their role are shown in table 4–1.

The data are proprietary to the SWG for a period of time for removal of instrumental effects and calibration. Formal data products are planned for release to the public. The products will include the basic products of the level 0 to level 1 processing (removal of instrumental and environmental signatures) and will be available from the National Space Science Data Center (NSSDC) (see chapter 16) in Flexible Image Transport System (FITS) format. Information on the availability and procedure for ordering the formal data products will be announced by the NSSDC at the appropriate time. In addition, it is expected that NASA Headquarters will support a Guest Investigator program for further research and analysis using the archived COBE data and will issue announcements for this program as these plans mature.

Name	Affiliation	Special Role
J. C. Mather	GSFC	Project Scientist and FIRAS Principal Investigator
M. G. Hauser	GSFC	DIRBE Principal Investigator
G. F. Smoot	UC Berkeley	DMR Principal Investigator
C. L. Bennett	GSFC	DMR Deputy PI
N. W. Boggess	GSFC	Deputy Proj. Scientist
E. S. Cheng	GSFC	Deputy Proj. Scientist
E. Dwek	GSFC	
S. Gulkis	JPL	
M. A. Janssen	JPL	
T. Kelsall	GSFC	DIRBE Deputy PI
P. M. Lubin	UCSB	
S. S. Meyer	MIT	
S. H. Moseley	GSFC	
T. L. Murdock	Gen.Res.Corp.	
R. A. Shafer	GSFC	FIRAS Deputy PI
R. E. Silverberg	GSFC	
R. Weiss	MIT	Chairman of SWG
D. T. Wilkinson	Princeton	
E. L. Wright	UCLA	

Table 4–1: COBE Science Working Group

4–2 Contents of the Database

Each instrument plans to produce a number of data sets appropriate to its particular mode of operation and scientific purpose. The sets for each include both time and position-ordered data.

The data set descriptions given here are illustrative only, since the process of analyzing and calibrating the instruments has only recently begun. All are preliminary and are subject to change for technical or budgetary reasons. Detailed descriptions will be provided when the data sets themselves are released.

2.1 The Formal Data Sets

The DIRBE is planning one time-ordered and two position-ordered data sets for public release. The time-ordered file is expected to contain every observation made by the DIRBE with the intensities from each detector given in physical units, fully calibrated, along with additional data such as various environmental parameters, sources of cross-talk from other instruments, and selected spacecraft operational telemetry. This will be the master DIRBE archive. The two position-ordered archives are related, the second derived from the first. Each week, all the observations made that week of each pixel on the sky are averaged and written into a position-ordered file, the calibrated annual file. Included are average intensities, noise statistics, and averaged environmental parameters such as solar elongation. The second position-ordered data set is similar to the first, but the intensities in each band are interpolated to a standard solar elongation of $90°$.

For FIRAS, the time-ordered data sets are planned to consist of the raw interferograms coadded in each pixel over time intervals corresponding to the interval between each set of calibrations, which are typically one month apart, and a corresponding file of the calibration model solutions for each set of calibrations. For pixels near the ecliptic plane, there are two entries in the interferogram file, each six months apart. For pixels near the ecliptic poles, there will be several coadded interferograms, typically a month apart. In addition, there are planned sky maps, which are position-ordered fully calibrated spectra obtained by applying the appropriate calibration models to the interferograms and averaging the resulting spectra for each pixel. These maps come in high, low, and uniformly spaced spectral resolutions.

The DMR time-ordered data set is planned to contain every observation by each of the six DMR radiometers, calibrations, and associated quantities such as attitude, instrument performance parameters, and environmental information. The position-ordered data set contains radiometric temperatures for each pixel of the sky.

2.2 Additional Data Sets

Additional data sets are planned for inclusion in the data archives as further analyses are completed. Such sets would include the results of application of astrophysical models to the data. For example, current plans for the DIRBE data include several models of the interplanetary dust scattering and emission, maps of the sky with the interplanetary dust contribution removed, and maps of the interstellar dust properties (such as column density, gas-to-dust ratios, effective dust temperatures).

Similarly, additional data products for FIRAS could include models (including effects of galactic light and dust) of the spectra for each pixel and maps of the deviations of the spectrum (corrected for the models and for the dipole anisotropy) from the average black body spectrum.

For DMR additional products might consist of the results of the spherical harmonic analysis of the sky, including the coefficients and the associated errors.

2.3 Archives Planned for the Guest Investigator Facility

Other data planned to be available in the guest facility include the spacecraft, mission planning, and attitude and orbit archives as well as many of the data sets used in producing the COBE data products. In addition, many other astronomical data sets (such as catalogs and reddening and radio maps) are planned to be reconfigured to the COBE coordinate system and geometry for direct comparison to COBE data.

An explanatory supplement, giving important documentation for the use of COBE data, is planned. It would include pertinent information about the mission, the method of processing the data, descriptions of end products, and an evaluation of the products, and would be distributed by the NSSDC.

4–3 Organization of the Data

The time-ordered data are based on the COBE telemetry format, which has a major frame of 32 seconds containing 128 minor frames per major frame. Each record typically contains all the data for the major frame.

The position-ordered data are organized by pixel number on the sky. The COBE is an all sky mission. Great attention has been paid to the coordinate systems, geometric projections, and storage structure for the map data. The ecliptic coordinate system (J2000) is the fundamental COBE coordinate system. The geometric projection chosen is the quadrilateralized spherical cube (Chan & O'Neill, 1975 and O'Neill & Laubscher, 1975). The celestial sphere is projected onto the six faces of a cube in a tangent plane projection. The lines of latitude and longitude on each face are then curved such that, when the cube face is divided into equally spaced rows and columns of pixels, each pixel is equal in area on the sky to every other pixel, and thus the maps preserve photometric integrity. The changes in the lines of latitude and longitude result in only minimal distortion of the shape of objects on a single face. The DMR and FIRAS use cube faces of 32 by 32 pixels, while DIRBE has 64 times that many (256 by 256). The pixels are numbered using a quad-tree structure. The basic sky maps are stored in order of pixel number, which means they are not rasterized. The quad-tree structure is a nearest neighbor scheme, and is hierarchical. For example, all 64 DIRBE pixels contained in a FIRAS/DMR pixel have consecutive pixel numbers and are stored contiguously.

In addition, selected sky maps should be available in rasterized form in traditional FITS image format, in Aitoff and possibly other projections, to facilitate easy display.

4-4 Future Plans

The COBE mission is still performing its all-sky survey and the data processing and analysis have just recently begun. At this time, plans for the data sets and for the guest investigator program are subject to significant revision. Details will be made public as plans become firm.

References

[1] Gulkis S., Lubin P.M., Meyer S.S., Silverberg R.F. (1990), The Cosmic Background Explorer Satellite, *Scientific American*, **262**, 1, p. 122.

[2] Mather J.C., Cheng E.S., Shafer R.A., Bennett C.L., Boggess N.W., Dwek E., Hauser M.G., Kelsall T., Moseley S.H. Jr., Silverberg R.F. (1990), A Preliminary Measurement of the Cosmic Microwave Background Spectrum by the Cosmic Background Explorer Satellite, *Ap. J. Letters*, **354**, p. L37.

[3] F. K. Chan & E. M. O'Neill (1975), *Feasibility Study of a Quadrilateralized Spherical Cube Earth Data Base*, Computer Sciences Corporation, EPRF Technical Report 2-75 (CSC), Prepared for the Environmental Prediction Research Facility, Monterey, California.

[4] E. M. O'Neill & R. E. Laubscher (1975), *Extended Studies of a Quadrilateralized Spherical Cube Earth Data Base*, Computer Sciences Corporation, NEPRF Technical Report 3-76 (CSC), Prepared for the Naval Environmental Prediction Research Facility, Monterey, California.

5

The many faces of the Archive of the International Ultraviolet Explorer satellite

[1]Willem Wamsteker
ESA IUE Observatory
P.O. Box 50727
28080 Madrid, Spain

5–1 Introduction

Since the archive of data obtained with the International Ultraviolet Explorer satellite is the first modern astronomical data archive it is of interest to follow its development. It is also the first heavily used archive of astronomical data. The IUE Archive has been from the beginning a distributed archive with three major archival sites, each one hosting a complete copy of the IUE Archive Data. Comparison of these allows one to evaluate how different design philosophies have affected the organization of the archive access and support. Over the past 12 years the linking of computers through networks has had a major impact, not only on the methods used for the maintenance of the IUE Archive itself, but also on its usage and accessibility.

5–2 The IUE project

2.1 Organisation and goals

The International Ultraviolet Explorer (IUE) satellite (Boggess et al., 1989; Faelker et al., 1989) was launched on January 26, 1978 as a joint project between NASA, ESA and SERC. The 3-axis stabilized spacecraft is equipped

---
[1] Affiliated with the Astrophysics Division, Space Sciences Department, ESTEC, Noordwijk, the Netherlands

with a 45cm Cassegrain telescope under normal incidence and 2 spectrographs with 4 cameras (single redundancy for each spectrograph) dedicated to spectroscopy of astronomical sources in the ultraviolet wavelength range extending from 1150 Å to 3200 Å. These spectra can be obtained in two resolutions: Low at $\lambda/\Delta\lambda = 300$ and High resolution at $\lambda/\Delta\lambda = 2 \times 10^4$. Each spectrograph covers half the wavelength range with a small overlap (short wavelength —Camera SWP— from 1150 Å to 1975 Å and long wavelength —Cameras LWR & LWP— from 1800 Å to 3200 Å). The high resolution spectra are obtained with an echelle optical arrangement and fill the full camera face of 768×768 pixels (8-bit depth) in a single exposure. In low resolution, the spectra cover only one third of the camera image. The brightness of the objects studied extends over a range of 10^9. The SEC cameras have only limited dynamic range (signal-to-noise ratio ~ 20) but extended brightness range can be easily obtained through variations in the duration of the exposures since the satellite is in geosynchronous orbit and in continuous contact with one of the two IUE Observatories (at GSFC for NASA and near Madrid at VILSPA for ESA).

The observing time is divided in two parts, one third for ESA and SERC and two thirds for NASA (8 hours and 16 hours respectively, each day). The science program of the IUE project is defined through an annual open call for proposals and time is granted to Guest Observers (GO) via a peer review process.

5–3 The IUE Archives

3.1 Location and Contents

Since proprietary rights on IUE data are limited to a period of six months, the need for a data archive was already recognized early in the project definition phase. Each of the three agencies involved made its own archival center to serve its own IUE users community (Giaretta et al., 1989). The NASA IUE archive is located at the National Space Science Data Center (NSSDC) at GSFC; the ESA IUE archive is located at the ESA IUE Observatory at VILSPA and the SERC IUE Archive is located at RAL.

Each of these three centres maintains a complete up to date copy of the IUE Archive. The IUE archive is currently defined as the contents of the Archive tapes (these are a copy of the data products supplied to the Guest Observer, usually within 24-72 hours after termination of his observations). The software used to produce the reduced data which are entered in the archive is IUESIPS (Turnrose & Thompson, 1987); the two observatories for NASA and ESA, each reduce their own data. Since the interchangeability of data taken at different observatories is vital for the value of the archive, meticulous care has been taken to document the development and maintain output compatibility of the IUESIPS software. The distributed nature of the IUE Archive has had strong influence on the ways in which each centre has developed its services over the years. Although the NSSDC Archive does contain certain specific datasets, other than science data, that have been obtained with the IUE spacecraft, we will not deal with them here. The current contents of the Science data in the IUE Archive after 12 years of orbital operations are given in table 5–1.

All data	Storage (Mbytes)	# of items	Total (Gbytes)
Raw image	0.59	78,030	46.0
Photom. corrected	1.18	78,030	92.1
Label Information	0.009	78,030	0.7
Low Resolution			
Line by Line spectrum	0.67	51,715	34.6
Merged spectrum	0.015	51,715	0.8
High Resolution			
Merged SW spectrum	0.66	12,817	8.5
Merged LW spectrum	0.74	12,840	9.5
Total		78,030	192.2 Gbytes

Table 5–1: Contents of the IUE Archive as of August 1990.

The current storage medium for the IUE Archive is magnetic tapes. However, with the advances of technological capabilities, different subsets of the IUE Archive have been created on various higher density storage media. To mention the most important, there are the extracted data subset at NASA's RDAF's (Regional Data Analysis Facilities at GSFC, Md. and at the University of Colorado, Co., which have been set up as reduction and analysis centers for archival research on spacecraft data) and at the Dominion Astrophysical Observatory (REF); the line-by-line spectra at ESA's ST-ECF (Pasian et al., 1990); and the low resolution spectra in the ULDA/USSP National Hosts (Wamsteker et al., 1989). Some of these will be discussed in more detail below.

3.2 General considerations

We will try to highlight here some aspects of the IUE Archive, which have been important for the development. It might be worthwhile to mention here also that the multinational nature of the IUE Project has also had strong influence on the developments in the Project and especially on its Archive. It is obvious that in a number of ways the IUE Archive is unique. Not unique in the sense that its model cannot be applied elsewhere, but in the fact that it was developed in a way which has anticipated the usage. In other words: the developments in the IUE Archive were such that all users could *directly* benefit from them. This contrasts to many other activities in this field where usually an archival planning is made either long in advance of the existence of the Archive (*e.g.* HST Archive), or it is done completely independent of the data acquisition, after the contents of the Archive have become fully invariable, such as the EXOSAT Archive.

There are a number of reasons why the IUE Archive has been exceptionally successful. I will try to summarize those here since they are in my opinion critical for any astronomical archive:

1. The Archive consists of both raw data and *data in physically meaningful units*, which can be directly applied to science problems.
2. The Project has from the very beginning striven to maintain control over processing and calibrations, ensuring the *homogeneity* of the archive contents.
3. The Archive has been built up —due to the short proprietary period associated with IUE data— with the intention that it should be able to support future observations.
4. Access and development have been driven by the Project science staff, which represents and is part of the user community.
5. Due to economic limitations the IUE Archive had to remain compatible with the technological capabilities of the user's institutes.
6. Although formally an archive was defined before the launch of IUE, the IUE Archive, as it stands now, has had a slow growth (driven by the acquisition of some 7,000 new data per year) which has allowed a very successful match between the Archive support needs from the project itself, the general infrastructure in Astrophysics, and the needs for the data by the users.

Aside from the above I would like to mention some aspects of Archives in general before getting to the details of the IUE Archive. Contrary to a common perception, the only reason for archives is to retain the availability of previously acquired information for a time when the increase of the cumulative knowledge of mankind justifies a re-analysis of the information contained in the archive. Although this is a relatively complex sentence, it is vital for those who consider to create new archives to understand the implications of the above. Often one sees confusion between an archive and a Database. The first has a use projected into the future, as well as a current value, while a database has no long term value although it may have an extremely high current validity. One could compare the two with a Library and a Bookstore. For these reasons the organisation of an archive is always a much more complex activity than that of a database.

One could typically consider that the catalogue of an archive —in the IUE case: the IUE Merged Log— is a database, *i.e.* the Catalogue is a part of the archive, but does not have the historical weight, since it could be "easily" reconstructed. On the other hand one should never forget that the catalogue design determines the use of the archive, since, if the catalogue is not well prepared, matched to the contents of the archive, and tailored to the current needs of the users, it will discourage even the most persistent scientist.

As a final comment I would like to point out that for all archives the two contradictory requirements of ease of access and educated usage represent an extremely complex issue. On one hand the requirement to spread knowledge stipulates completely free access to the contents while on the other hand, the data are bound to be modified, destroyed and/or abused unless they are carefully safeguarded. It is one of the advantages of the present times (easy duplication and electronic access) that one can actually begin to make an attempt to fulfil these two functions simultaneously.

3.3 Development history

The evolution of the IUE Archive can very simply be summarized by realizing that in 1978 its contents were 0.0 bytes and in 1990 the contents amounted to 190 Gbyte while between these two dates an archive retrieval of 223,250 spectra ($\sim$ 250 Gbyte) has been supported. These numbers suggest that,

1. the archive usage is strongly driven by the availability of new data and,
2. the needs for data distribution have far outstripped any originally foreseen capabilities.

The historical evolution of the IUE Archive is best seen following the highlights given in table 5–2, which identifies the most important developments related to the IUE Archives at the different Archive centres. Looking at table 5–2 it is clear that the multi-agency nature has had a strong influence on the archive development. Since each centre had to accommodate the needs of its own current IUE users community, different centres have responded in culture-typical ways to this challenge.

Table 5–2 clearly shows that at the start (1978) the IUE Archive was a more or less classical data archive involving a large amount of human

	Milestone	Archival site [a]
1978:	Creation of IUE Archive;	NSSDC/VILSPA/ RAL/GSFC
	Start data entry;	GSFC/VILSPA
	Catalogue Distribution: printed	GSFC/VILSPA
	Data request distributed by mail (tapes)	NSSDC/VILSPA/RAL
1979:	Creation of Merged Log	
	Distribution: Tape & Microfiche	NSSDC/VILSPA
	Printed	GSFC
1981:	RDAF and CURDAF creation	GSFC/COLORADO
	IUE Merged Log distribution (tapes)	CDS
1982:	Extracted spectra on mass storage	NSSDC/GSFC
	IUE Catalogue remote access (UK only)	RAL
	Extracted spectra distribution to RDAF's	NSSDC
	Data requests filled locally (tapes)	RDAF's
1984:	Catalogue remotely accessible (Worldwide)	VILSPA/NSSDC
	Distribution of low resolution extracted spectra on microfiche	RAL
	Entry of remote dearchiving requests	VILSPA
	Running of remote dearchiving jobs	RAL
1986:	Local browse file (photographic)	GSFC
1987:	Entry of remote dearchiving requests	NSSDC
	Data copying via network	NSSDC
	Homogenization of Object Names	VILSPA/CDS
1988:	ULDA/USSP version 1.0 distributed	VILSPA
	Monthly IUE Merged Log exchange between the two Observatories via SPAN	VILSPA/NSSDC
	Extracted spectra to DAO	NSSDC
1989:	Line-by-line data loaded on optical disk	ESA-ST/ECF
	IUE Merged Log accessible through ESIS	VILSPA/ESRIN
1990:	All raw data loaded on optical disk for Final Archive processing	NSSDC
1994?:	All data available through remote access	???????

Table 5–2: Milestones in the access, use and maintenance of the IUE Archive.

[a] In this table we have separated the activities associated with the NASA IUE Observatory (indicated as GSFC). We have included also some activities which are not directly associated with the IUE Project, but have been part of the general developments to facilitate the access to the IUE Archive, *i.e.* the distribution of the data content of the Archive and the associated primary catalogue. An example of such activities is the work of CDS and others as indicated. Most of the developments done at RAL were done under the STARLINK Project.

interventions (tape handling) and paper bookkeeping to fulfil the demand of the users. The obvious tasks required included keeping track of IUESIPS changes (many in the early days), informing the community on the contents of the rapidly growing archive, controlling proprietary rights, synchronizing archive buildup in the different archival sites and many others. This resulted in a number of difficulties to maintain the archive, which stimulated the gradual process of archive development. No major change in the IUE Archive itself occurred until 1981, when the extracted spectra were loaded on the IBM mass store at GSFC. This step made in 1982, has been one of the most critical developments, since it created a mechanism to handle the ever growing data archive more efficiently at a time when the size of the Archive was still manageable with the technology available. Looking at table 5–3 it is clear that support for dearchiving is coming mainly (80%) from the ULDA/USSP, the RDAF facility and the network dearchiving at NSSDC. None of these solutions would have come about if the extracted data set creation, needed for the mass store, would not have occurred.

Although in the interval between 1978 and 1982 no changes were made in the data distribution mechanism, many steps were taken to improve the maintainability of this distributed archive and to keep the data accessible. Although the most visible activities are indicated in table 5–2, I would like to comment that the creation of a common database handling agreement of which the IUE Merged Log is the visible part, has been one of the most important —but difficult to accomplish— tasks.

The next major change I would like to highlight is the remote access to the Catalogue (IUE Merged Log) which was brought on line by RAL in 1982 for the UK. This allowed for the first time rapid and convenient visibility of the contents of the IUE archive to all UK astronomers who could make use of STARLINK facilities, without all the troubles that go along with keeping such dynamic general information on one's own systems. The rest of the world was supplied with a similar facility in 1984 when both VILSPA and NSSDC brought the IUE Merged Log on line through more general network connections. At RAL and NSSDC these on-line catalogues were brought up under self-developed search packages. At VILSPA a commercial database was chosen since it would allow at the same time a great simplification of maintaining the archive, as well as manage the various data exchange controls which had to be kept under control. At present, archive maintenance is fully under the same database system, with an SQL layer added.

Various subsets of the IUE archive were maintained in tape form for possible future use at all three centres. In 1984 both RAL and VILSPA introduced support for remote dearchiving requests. There exists a clear difference in the way these were implemented. At RAL, the relatively homogeneous computer environment for the full UK community allowed the remote user to start the actual dearchiving jobs, the tapes so created being sent by mail immediately afterwards. At VILSPA, only the request for data could be filed. This was then further handled in the classical way except for the fact that all information needed for the actual dearchiving activity is easily generated from the master archive database. In 1987, the NSSDC took the next natural step and allowed dearchiving requests to be filed, and then processed from the mass store, bringing the data up front to a disk space accessible to the user, who can copy the data to his/her home institute.

An important catalogue development was started in 1987 by CDS and VILSPA to homogenize the object names for all observations in the IUE archive under a rudimental hierarchy defined by the IUE Project.

In 1988 a fully new concept was introduced by VILSPA with the first release of the Uniform Low Dispersion Archive (ULDA) with its associated Support Software Package (USSP). As the name says, this concerns only low dispersion spectra directly suitable for scientific analysis (data in physical units only). This concept took the distributed archive one step further by creating a set of national host institutes (one per country: see table 5–4) each of which was supplied with this controlled subset of the archive, the ULDA. The software supplied consists of two parts: one running at the national host (selection program QUEST) and one running at the user's institute (Unscrambler UNSPL). The main functions of the two being, respectively, (a) to select the data, and (b) to organize their transfer and allow their formatting in whatever favourite data analysis package is wanted (*no special project related format for the user to worry about*). One of the main advantages of this system is that the data have been reformatted to minimize disk space usage ($\sim$ 80 Mbyte for 42,000 spectra in ULDA version 3.0). Also the actual process of dearchiving involves only the user of the data, while the software maintains its own disk usage (*i.e.* essentially maintenance free after the installation at the Host and distribution of the unscrambler to the users). Associated with the ULDA/USSP distribution is the production of a series of scientifically and subject oriented ULDA Guides (La Dous, 1989; Festou, 1990), which help the novice scientist in his first approach to the massive amount of data collected in the IUE Archive.

The next major step will be the creation of the complete newly reduced IUE data archive on optical disks, the IUE Final Archive (IUEFA). The IUE Final Archive will give a whole new dimension to the possible use of the IUE data archive, but I suspect that the modes of access to the data will remain similar to those currently in use. They seem to have reached an efficiency which can barely be improved beyond the present 5 spectra per hour currently retrieved from the IUE Archive, unless drastically different philosophies are applied to the overall system. In view of the capability of individual scientists to analyse data, this appears unlikely.

Let me finish with reiterating the clear difference between the way the modernization has been driven in the US and in Europe. Although the functionality of both approaches is obviously very high (see table 5–3) the major emphasis in the USA has been to bring the user to the data while in Europe the emphasis both in ESA and SERC has been to bring the data to the user. As such the currently most efficient means to access the data in the US (RDAF facilities) and in Europe (ULDA/USSP) clearly illustrate this. It must be pointed out that arrangements like the ULDA are not trivial to achieve when applied to the transfer of high resolution data with the current magnetic disk technology. Also a major redesign of QUEST would be required to meet the specific scientific requirements for efficient access to high resolution data. In that respect the RDAF's present a more general solution for the archive as a whole. On the other hand solutions like the RDAF's require long term commitments by funding Agencies.

Although other approaches to data distribution than the ones chosen by the IUE Project can be imagined, the extraordinary high usage, especially over the last 2 years (see table 5–3) clearly illustrates that approach of the IUE Project —essentially supporting the data distribution in a *"get what you want, but no more than you want, when you want it"* mode— has caught on very well with its user community, and has allowed this community even to grow far beyond the original IUE users. It may be quite possible that this could also be the most cost-effective, and from the user viewpoint, the most convenient solution in general.

3.4 Usage of the IUE Archives

Due to the very general applicability of the data of IUE to nearly all astrophysical problems, it is difficult to describe the general user of the Archive. Most astronomers currently active in research and education have been at one stage or another been directly involved with IUE data over the last 12 years. One special class of astronomers can be clearly identified as heavy users of the IUE data: those, whose interests are strongly oriented towards the behaviour of astronomical objects in more that one single wavelength interval, multi-wavelengths pundits.

Table 5–3 shows that the developments, which annually support most of the dearchiving of IUE data are the ULDA/USSP (58%) and the RDAF's (22%).

The large usage supports mainly current research. This is illustrated by the fact that most proposals for IUE observations are made after a careful

	Total No. of spectra	Annual rate	% annual	% total
12 yrs VILSPA-Tape	34,600	2,880	6.8	15.5
2 yrs VILSPA-ULDA/USSP	49,000	24,500	58.3	21.9
12 yrs NSSDC-Tape	21,825	1,820	4.3	9.8
3 yrs NSSDC-Network	6,025	2,010	4.8	2.7
12 yrs RAL	19,800	1,650	3.9	8.9
10 yrs RDAF	72,000	7,200	17.1	32.3
10 yrs CURDAF	20,000	2,000	4.8	9.0
Total	223,250	42,060		

Table 5–3: Usage statistics of the IUE Archives (August, 1990)

analysis of the data in the archive and supported by the large quantity of publications based on IUE data (1,700 papers in the four major astrophysics journals; *see* Mead *et al.*, 1986 for a complete bibliographical index from 1978-1985). This is unmatched by any other facility in Astronomy. Also the large number of Ph.D. theses based on IUE results —often from the archive— suggest that one complete generation of young scientists has been strongly influenced by the IUE Project and its Archive.

3.5 Future of the IUE Archive

At the present time the IUE Project is preparing for a major activity, still to be done during the operational life of the satellite: *the preparation for the final Archive*.

This is the first time that such an activity is planned to be finished before the termination of the orbital operations of the satellite itself for any space science project. After about three years of preparation the current planning foresees to reprocess all IUE data with a single new Software package (NEWSIPS) developed under the Project by NASA. There is a fundamental difference between the design philosophy of this and the original IUESIPS: while the last was designed to supply reduced data to the Guest Observers, the second is designed to optimize the existing data and to create a homogeneous archive of the highest possible quality [8]. The project is supported in this activity by its users community through the NASA-FADC under the inspiring chairmanship of Prof. J. Linsky. Let us hope that this ambitious plan will be brought to a good end and that, as suggested in table 5–2, a modern and homogenized IUE Archive will be ready on schedule and remain as useful as the current IUE Archive. Although it has been established that this archive will be optical disk resident we will still have to face the important task of defining the associated database (catalogue) and the access mechanisms, which should remain tuned to its users rather than to reasons extraneous to scientific research. Only this way the productivity of IUE will remain, even long after the spacecraft has stopped sending data down to Earth.

More detailed information on the access and the contents of the IUE Archive can be found in the IUE Newsletters (especially Grady and Taylor, 1989 for NASA; Driessen and Pasian, 1988, and Barylak, 1988 for ESA). These are published quarterly by the ESA and NASA IUE Observatories. A complete in-depth history of the IUE Archive is recorded in the minutes of the IUE 3–Agency Coordination meetings, held twice a year, and published by the hosting Agency with a limited circulation.

References

[1] Barylak M. (1988), *ESA IUE Newsletter*, **31**, 41.
[2] Boggess A., Wilson R., Barker P. J., Meredith L. M. (1989), in *Exploring the Universe with the IUE Satellite*, ASSL **129**, Y.Kondo (Ed.), Kluwer Acad. Publ., Dordrecht, 3.
[3] Driessen C. & Pasian F. (1988), *ESA IUE Newsletter*, **30**.
[4] Faelker J., Gordon F., Sandford M. C. W. (1989), in *Exploring the Universe with the IUE Satellite*, ASSL **129**, Y.Kondo (Ed.), Kluwer Acad. Publ., Dordrecht, 21.
[5] Festou M. (1990), IUE-ULDA Access Guide **2**: Comets, ESA SP-1134, W. Wamsteker (Ed.).
[6] Giaretta D., Mead J. M., Benvenuti P. (1989), in *Exploring the Universe with the IUE Satellite*, ASSL **129**, Y. Kondo (Ed.), Kluwer Acad. Publishers, Dordrecht, 759.
[7] Grady C. A. & Taylor M. A. (1989), IUE Data Analysis Guide, NASA IUE Newsletter **39**.
[8] IUE_DCG (1990), Reference Document of the IUE Final Archive.
[9] LaDous C. (1989), IUE-ULDA Access Guide **1**: Dwarf Novae and Nova-like stars, ESA SP-1114, W. Wamsteker (Ed.).
[10] Mead J. M., Brotzmann L. E., Kondo Y. (1986), *NASA IUE Newsletter*, **30**.
[11] Pasian F. *et al.* (1990), ESA SP-310, *in press*.
[12] Turnrose B. E. & Thompson R. W. (1987), IUESIPS Information Manual, Version 2.0 (European Edition; J. Clavel).
[13] Wamsteker W., Driessen C., Munoz J. R., Hassall B. J. M., Pasian F., Barylak M., Russo G., Egret D., Murray J., Talavera A., Heck A.(1989), *Astron. Astrophys. Suppl. Ser.*, **79**, 1.

Main centres

NSSDC :	Site Manager: `NSSDC::PERRY`
VILSPA :	Captive Account `28843::VILSPA` (Password: `DB`)
RAL :	STARLINK member authorized access only. `19457` (SPAN)
RDAF/GSFC :	Site Manager: `IUE::RTHOMPSON`
CURDAF :	Site Manager: `CYGNUS::BRUGEL`

ULDA National Host Managers:

Belgium :	Fax :	+32-2-3749822
Brazil :	Span :	`47556::LUIS`
Canada :	Span :	`NCF::PSI%DAO::CRABTREE`
		`telnet%nootka.dao.nrc.ca::crabtree`
China :	Tel. :	+86-551-331760
France :	Span :	`CDSXB2::JASNIEWICZ`
Germany :	Span :	`AITMVX::SCALES`
Holland :	Earn :	`VERBUNT@HUTRUUS10`
Italy :	Span :	`40057::FRANCHINI`
Spain :	Span :	`28843::CD`
	Earn :	`IUEHOT@DDAESA10`
Sweden :	Span :	`21619::LABAN::KE`
(The Swedish centre serves also Denmark, Norway and Finland)		
Switzerland:	Span :	`20579::UGOBS::LANZ`
ST–ECF :	Span :	`ESO::PASIAN`
Taiwan :	Span :	`PSI%0487230035::WHSUN`
	Bitnet:	`NCUT@TWNMOE10`
U.K. :	Span :	`19457::JM`
	Janet :	`UK.AC.RL.STAR`
USA :	Span :	`NCF::ULDA`

Table 5–4: IUE Archive access information

6

Data Archive Systems for the Hubble Space Telescope

---

[1]E. Schreier, [2]P. Benvenuti & [2]F. Pasian

[1] *Space Telescope Science Institute*
3700 San Martin Drive
Baltimore, MD 21218, U.S.A.

[2] *Space Telescope – European Coordinating Facility*
c/o European Southern Observatory (ESO)
Karl–Schwarzschild–Str. 2
D – 8046, Garching bei München, FRG

6–1 Introduction

The Hubble Space Telescope (HST) has been successfully launched in April 1990. In spite of the known problems due to the spherical aberration in its primary mirror, and from the analysis of the data collected up to now (November 1990), it appears that HST will indeed be the source of unique and outstanding science for the whole duration of its mission. The oversubscription rate for HST observing time is such that the number of accepted observation proposals is an order of magnitude lower than the total proposals. It is therefore essential for an efficient archive facility to be established to allow the world-wide astronomical community to share the tremendous scientific capabilities HST will offer.

The need for an archival system for HST has been recognized for a long time. As a long-lived mission, it was obvious that the archival data produced by HST and put in the public domain would eventually comprise an invaluable astronomical resource. It became also clear that the Science Operations Ground System (SOGS) provided only minimal capabilities for writing output data to tape, and no real catalog capability or data retrieval system. In 1982, the Space Telescope Science Institute (STScI) initiated discussions with the ST Project about the needs for an archival capability, and participated with it in discussions of how to remedy the situation, especially in regard to the requirement to transfer a copy of the data to the Space Telescope–European Coordinating Facility (ST–ECF), as per the NASA–ESA Memorandum of Understanding. However, no archival capability resulted from these early discussions. That year the STScI also proposed to NASA Headquarters (Information Systems Office of OSSA) to study optical disk technology for archiving astronomical data. The resulting grant allowed us to procure an optical disk and workstation, and to start understanding the specific requirements for performing optical disk archiving.

In 1983–84, the ST Project working with the Astrophysics Division at NASA Headquarters embarked on a major study of the requirements for an HST archive system which eventually lead to the Data Archive and Distribution System (DADS) procurement. The STScI was encouraged to actively participate in this study which resulted in a set of requirements and the initiation of a procurement. However, uncertainties as to available funding and prioritization of requirements led to delays in the procurement cycle and the certainty that DADS would not be available by launch time. This in turn led to a recommendation by the NASA–ESA Working Group that an interim system be developed by the STScI in collaboration with the ST–ECF. In response, the Data Management Facility (DMF) project was started in July 1985.

The DMF had the limited goals of: (1) recording the science data on optical disk; (2) producing a catalog of the data; and (3) making a copy of the data for the ST–ECF. Prototype capabilities were developed in time for the original launch date (December 1986). The first real use of DMF was made for the archiving of the Guide Star Selection System (GSSS) plate scans at the STScI (see chapter 7, this book) and of the IUE line-by-line images at the ST–ECF. The GSSS is a collection of about 6,000 magnetic tapes (800 Gbytes) containing a complete set of digitized scans of the sky which was transferred onto 500 optical disks, representing probably the first scientifically useful large-scale optical disk archive. The IUE Line-by-line Archive consists of about 23,000 image-strips of the Low Dispersion IUE spectra. The content of about 250 magnetic tapes (18 Gbytes) was transferred onto 10 optical disks and the archive was operationally tested in collaboration with the IUE–VILSPA Observatory.

The final decision to proceed with the DADS procurement was not made until mid-1987, following discussions with the Goddard Space Flight Center (GSFC) director as to where to physically locate the HST archive. Actual work on DADS development started only in August 1988. Given the delays in the DADS procurement, and the prospect of extended operation of DMF before DADS comes on-line, capabilities have gradually been added to DMF. At STScI, DMF now supports not only the HST data and guide star scan archives, but also the Calibration Data Base System (CDBS) for HST. In addition, an increasing amount of functionality has been added in order to carry out initial HST data operations.

In the following sections we first discuss the design issues and the required capabilities for an HST Archive System and then we briefly describe the current implementation of the DMF System.

6–2 Critical design issues for the HST Archive System

There are three major external factors that directly influence the design of the archive system for HST data: the nominal 15 year timeline for the project; the volume of data to be managed; and the environment in which the system must operate.

The long timeline is an issue for a number of reasons, including:
- expected changes from second generation instrumentation;
- expected changes in system use, driven by science needs;
- expected changes in technology, and technological obsolescence of hardware platforms, operating systems, protocols, and storage devices; and,
- associated costs of system maintenance and enhancement.

The volumes of data to be archived are just now within the grasp of current technology, but only if a limited amount of data is stored on-line.

The data volumes are an issue because of:
- the desire to have more than three years of data on-line;
- the significant amount of planned data to be stored;
- possible volume changes driven by:
    1. new instruments;
    2. new processing algorithms;
    3. additional data products to be archived;
    4. archiving of new Guide Star scans and products; and,
    5. proposed extensions to new kinds of data (software, scanned documents, *etc.*).

The third critical factor is related to the environment in which DADS must run:
- heterogeneous in both systems (VMS & UNIX) and protocols (TCP/IP & DECnet, OSI in the future);
- requiring support for operational and science users, novice and experienced users;

- requiring support for random local and remote data access, and for bulk distribution to archive sites; and,
- requiring support for automated ingest from ground segment operations (SOGS), manual ingest from the calibration system (CDBS), and bulk ingest of optical disks (GSSS plate scans).

2.1 Derived critical design issues

The external factors described above are requirements on DADS that are mostly reflected in the existing system specifications. From these system requirements, it is possible to derive a set of requirements on the design of DADS itself. If DADS was just to be a short-lived system, and only needed to operate in a constrained environment, then an easy, 'point' solution might be acceptable. Such solutions are cost-effective for limited, short-term, projects because the development costs are lower and the window of exposure to maintenance problems is reduced. For a long-term project like DADS such an approach is totally unacceptable because the up-front savings in the development cycle are quickly overtaken by maintenance and enhancement costs.

Reformulating these system requirements in terms of a set of critical design issues, we arrive at the following:

Extensible/Evolvable Design: Because of the lifetime of the project it is critical that the design of the system be extensible (second generation instruments, new data types, new uses) and evolvable (new hardware, operating systems changes, new interfaces, new technology). Clean isolation of interfaces and system or device dependencies, sensible layering of the system, use of appropriate language and build tools all are essential here.

Maintainable Design: The system will have to gracefully survive minor environmental changes (operating system revisions, compiler versions, new interface ROMs) and also minor changes in use (new keywords, altered data formats, new access mechanisms). Each of these items should ideally require little more than a change to a table entry, or a simple re-compile/re-link to accommodate. Maintainability also requires good modularity, functional strength, good documentation, and adherence to other coding disciplines. Use of an appropriate computer language (*e.g.* C language) is part of this consideration, as is isolation of system and device dependencies.

Data Volume Issues: Because data volumes are likely to increase over time, and storage technology will also evolve, the requirement to be able to accommodate change in this element of the system is essential. Optical disk technology seems to be changing on four to five year cycles and this is likely to continue. New technology with even higher data densities, higher data rates, and lower costs can be anticipated. Designing the system, interfaces, and data handling mechanisms so that changes to these critical elements can be sensibly accommodated will require careful attention to the system layering, modularity, and interface isolation considerations already mentioned.

Heterogeneous Environment Issues: The diversity of the environment is a critical issue, that must be faced from the outset. Part of the solution for this problem is a clean design of interfaces, as already mentioned above. The need to support a heterogeneous environment imposes additional considerations on where interfaces must be defined, and adds fundamental considerations of data storage modes, and data management and conversion. Where diversity already exists, it will likely increase over time, *e.g.* adding OSI networks, new optical disk types, or new hardware architectures and word formats. Failure to properly handle these issues can lead to significant maintenance and enhancement impacts over time.

Access and Security Issues: The design considerations that derive from the access and security issues also require careful attention to system interface definition. DADS has an interface to the Science Operations Ground System and this must be highly secure so as to avoid any possibility of affecting the operational environment. Furthermore, DADS must protect proprietary data, while at the same time serving users in the open world of international networking. Approaches to security are dependent upon the particular implementation strategies chosen, so that only general statements can be made at this point. Mechanisms for dealing with the diversity of user requirements, and with a wide-spread user community, must be integrated into the system.

Flexibility and Adaptability Issues: The last design consideration is a more general one, and also more difficult to define in terms of design rules, though it underlies the other issues. It is a given that we will want the system to do things that it was not exactly designed to do (new data, new access modes). It is also a given that we will need to change the archive system over its lifetime (for all of the reasons discussed above). In order to accommodate this requirement gracefully, the system must be well designed and well structured, and the designer must have seriously considered the issues of adaptability during the design process. Putting the "hooks" in, defining appropriate interface protocols, foreseeing the future uses, and configuring things so that change may be accommodated are hallmarks of good design and are almost impossible to legislate, but the absence will kill long term system viability.

6–3 The Capabilities Required of the HST Archive System

The data volumes produced by the HST mission will be prodigious, with several gigabytes per day needing to be archived. A wide variety of data types are involved including spectroscopic, imaging, photometric and astrometric scientific data, engineering, ephemeris and scheduling files. The rôle of the ST archive system is to manage this enormous amount of data and to organize them so that users can efficiently find the information they need.

There are three basic areas that the system must address: accepting and storing the data for ingest, managing the user's access to the data including protecting the proprietary rights on scientific data, and distributing the data.

The primary sources of data are the SOGS and CDBS systems which provide science and engineering, and calibration data respectively, and Guide Star Plate Scans. For SOGS and the CDBS the archive is the end destination of most of their products. As the ground system matures, the archive will also support many other users although they will probably not contribute as great a volume. Instrument scientists monitoring the health of the spacecraft will store trend information in the system. Value added information derived from ST observations may be put into the archive which will be the central depository for all ST related information.

Two distinct groups of users need to be served: operational users for whom the archive is a resource needed in the daily activities supporting the telescope, and scientific users, who will use the archive to get data to be analyzed for astronomical purposes. Because of security concerns, no interactive interface will be supported between SOGS and the archive, but full capabilities to archive and retrieve files must be provided. Operational users on the science cluster, *e.g.* CDBS or instrument scientists, will need a more flexible means of looking at the contents of the archive including interactive network connections. Astronomers will want a very responsive system which allows them to find information they need without any vast knowledge of the archive system or instruments.

To serve these interactive users, the archive system will need a sophisticated user interface, together with a catalog which will index the contents of the ST archive. The interface must hide the internal complexities of the archive, and guide the user as to how to get the data desired. The catalog is needed to enable the user to find specific datasets of interest without having to actually inspect the data itself. The catalog must also tie the contents of the archive together, indicating which files were used in the calibration of a given image, or vice versa, which images were calibrated using a given file. The catalog must answer such questions as what observations have been taken of a given target, by which instrument, within what interval, and also tell the user what is the appropriate ephemeris, calibration, *etc.*, for these observations. The user interface must show how to accomplish these queries, and how to get the requested files.

It is in the distribution of data that the heterogeneity of the user community comes most to the fore. The archive system must be able to serve the very diverse needs of users when it delivers the data. User at the STScI will need enormous volumes of information delivered electronically. The single largest data flow in DADS will be supporting distribution to other archive sites, the ST–ECF, the Canadian Astronomy Data Centre (CADC) (see chapter 20), and such other sites as are approved. Some external users will wish to get data delivered electronically over a variety of networks. Many users, at least for the present, will want their data on traditional magnetic tape. New media, such as cartridge tapes and small scale optical disk, may be very cost effective, and the archive system should be capable of accommodating them. The possibilities available for both electronic and hard media distribution are increasing enormously and the archive system will need to keep up with the demands of the user community.

6–4 The Data Management Facility (DMF)

As stated in the introduction, the Data Management Facility (DMF) is a system capable of ingesting HST data, recording them on optical disk, producing a catalog of the data and, at the STScI, making a copy of catalog and bulk data for the ST–ECF. Additional built-in capabilities allow DMF users to browse the HST catalog, to request data to be dearchived, and to receive data either via network or on hard media.

4.1 The DMF catalog

The HST catalog is the database that contains the information describing the files archived on optical disk by the DMF system. This database derives its information from data files passed to the DMF machine at STScI. The data files come from two primary sources: the science header files of SOGS data sets, and files containing information from the Proposal Entry Processing System (PEPSI). These files are archived and processed by the DMF Ingest utility, which processes their contents and places the relevant information in the DMF Catalog.

The HST catalog is handled by a relational database management system, and includes a large number of relations (or tables), *e.g.* tables related to the observations, to security, to calibration and engineering data, archive index tables, mission schedules and timelines, or business management and accounting tables (Silberberg *et al.*, 1989 [5]). There are 11 main relations which form the basis for the catalog of the HST science data: PROP, OBS, EXP, DSNAME, FILES, and 6 so-called master tables. The master tables contain detailed engineering and instrumental parameter values for each observation. There is a separate table for each instrument.

The PROP relation contains administrative information for each accepted HST proposal, such as the proposal title, the scientific category, and the Principal Investigator (PI) name and address for every accepted HST observing proposal. The OBS relation contains the proposed instrumental parameters for each HST observation, such as PEPSI proposal, exposure, and line numbers, target descriptors, target alias, instrumental configuration and operating mode, aperture, spectral elements, and exposure time. Both tables are populated with PEPSI information. The EXP relation, along with the master tables, contains the primary information about each completed HST science observation. This relation is populated by the DMF ingest process by extracting the field values from keywords in the headers of the science datasets as supplied by the Post Observation Data Processing System. The DSNAME relation is the basic index of every *dataset* archived onto DMF optical disks, while the FILES relation is the basic index of every *file* archived onto DMF optical disks. Each of the above tables, from PROP to FILES, is linked to the subsequent one by means of a 1-to-N relation.

The complexity of the HST catalog is notably increased by the large number of additional tables containing auxiliary information. Fortunately, the DMF user does not need to deal directly with this complexity. Even if each catalog table can still be individually accessed, the DMF user interface allows the definition of "views" on the database, which can span over different tables. Therefore, on a single screen, information extracted from different tables can be displayed, and search constraints can be specified. In the future, it will even be possible for a user to create his own personal HST catalog interface screens (Silberberg *et al.*, 1989 [6]).

4.2 The DMF archive

All data files related to HST observations are stored on optical disk. This includes raw data, calibrated data, engineering data, on-board computer dumps, real-time activity log, observatory monitoring system, *etc.* producing an average of 0.8–1 Gbyte a day. Calibration reference files are also stored on optical disk.

The size of the archive and its rate of growth force the data to be kept off line. Nevertheless, the above-described catalog structure and the DMF software allow data to be retrieved with jukebox and/or operator intervention with quite reasonable turn-around times (from some minutes to some hours, depending on the size of the request.)

DMF operations at the STScI are such that two copies of each optical disk, one for STScI and one for ST–ECF, are simultaneously made. The device selected for this purpose is the LMSI Laserdrive 1200, using double-faced media holding a total of 2 Gbytes per optical disk.

The expected lifetime of optical disk media is in the range of 15–20 years, and the lifetime of the HST archive is even longer. In such a long period of time, the developments in computer industry may follow directions totally unpredictable today. Therefore, although at the moment the optical disk managing system is run on specific hardware and on computers using the VMS operating system, provisions have been taken to allow media portability, in particular to different computer platforms. The physical structure of the optical disks and their format is such that they can be read in different environments, and tests to access them from Unix platforms have been already successfully carried out. The format of the optical disks is described in (McGlynn & Hunt, 1988).

4.3 The DMF software

The DMF software has been built as a joint effort of STScI and ST–ECF; it is composed of a user interface and of an archive subsystem. The overall structure of DMF has been described in detail elsewhere (McGlynn *et al.*, 1988). The DMF components and their inter-relations are shown in the figure. The user of the archive, after having accessed the catalog of the HST observations through Starcat, the user interface, issues a request to a Request Handling procedure. The request is queued to the File Handler main process and is serviced through the optical disk (OD), the jukebox (JB) and the receiver processes, allowing the user to get the data. The superdirectory (SD) process keeps on magnetic disk directories of the files contained on optical disk, and of the optical disks themselves. The Operator Interface Process (OIP) helps the operator to maintain the system. The dashed box encompasses the DMF archive subsystem.

The system is layered on some interface libraries, allowing portability among different database, networking, hardware, operating system environments with a limited amount of effort. *STDB* interacts with different database management systems, *NET* relates to the use of different networking protocols and allows communications among the various DMF processes; both have been developed at the STScI. *Proteus* is a general-

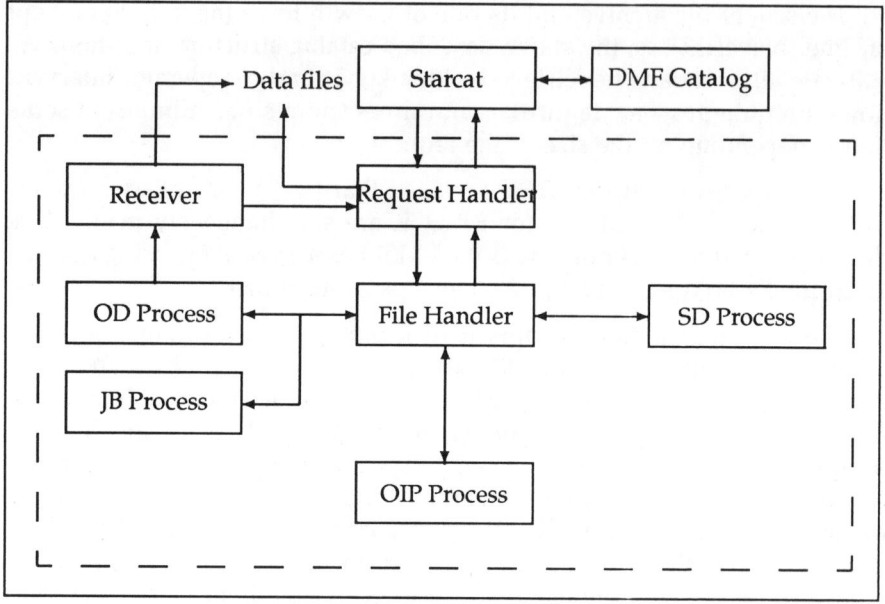

Figure 6–1: Structure of the DMF

ized application interface developed at ST–ECF and *TermWindows*, an interface to screen handling and help management, has been developed within ESO/Image Processing Group.

Each of the building blocks of DMF are briefly described in the following.

STARCAT is an acronym for Space Telescope ARchive and CATalogue, and is a program running on both VAX/VMS and Unix platforms. It has been implemented jointly by ST–ECF and ESO as a user interface to a set of services available to astronomers: access to astronomical catalogs, retrieval of files from the archive, access to remote databases and access to Astronews (ESA's Space Science Dept. Astrophysics Division bulletin board). Its main task is, of course, to provide access to the HST catalog and to allow users to request HST data to be dearchived. Users can browse through the catalog by means of various "forms" corresponding to catalog relations or views (*e.g.* proposals, observations, instruments, targets, *etc.*), specifying the search constraints, and "marking" the data of interest among the retrieved information. This selection operation builds a list of files to be dearchived.

STARCAT allows also access to astronomical catalogues (currently about 40), including the HST Guide Star Catalog, observation logs, a number of

Data Archive Systems for the Hubble Space Telescope

object-type oriented astronomical data catalogues; catalogues containing general information are also available. There are some other services available on the ST–ECF installation, such as connection to remote databases and access to a local news bulletin board, which includes HST and NASA news and IAU telegrams.

The archive subsystem, currently implemented on the VMS platform, is responsible for managing the archive and retrieval of datasets requested by users or other portions of the facility. This system takes care of all hardware-dependent issues involved with data storage, making the details of form, media and location of datasets invisible from the outside.

The main File Handler process is the core of the archive subsystem. This process handles all communications with users and other elements of the facility. It initiates subprocesses talking directly to optical disks and superdirectory tables, reads and analyzes requests for file handler resources, queues these requests among the components, and relays information of the eventual disposition of a request back to its originator.

The Optical Disk (OD) process is responsible for the actual access to optical disks and runs as a subprocess of the File Handler process. A number of OD processes may be active simultaneously: each of them is responsible for access to a single optical disk drive. The primary responsibilities of the OD processes are mounting and dismounting disks, and transferring files from and to optical disk, to and from magnetic disk or tape.

The superdirectory (SD) process is used to access and modify the superdirectory of the archive. The superdirectory is a magnetic disk file containing entries for each file on the optical disk archive; it is expected to eventually contain more than one million entries. Aside from the superdirectory itself, also a table of the disks belonging to the archive is maintained.

The Jukebox (JB) process is used when optical disks are to be moved in or out of a jukebox, imported or exported. There must be one JB process for each jukebox.

Operator interactions with DMF are mediated by the Operator Interface Process (OIP). This process receives requests and messages sent by other DMF processes and enables the operator to reply to these and to take appropriate actions. The actions could be checking the status of the system, adding or deleting disks from the archive superdirectory, mounting or dismounting optical disks, archiving/retrieving files from optical disks, moving a disk to/from a jukebox (if relevant), *etc.* The OIP provides a friendly interface, based on both prompt mode and menus and forms, allowing the operator to issue commands without bothering about their syntax, while critical operations (such as disk initialization, removal from superdirectory, *etc.*) have the appropriate security checks.

It is necessary, for the DMF system, to efficiently handle simultaneous dearchiving requests, queue them according to requestor and medium, dearchive data so as to minimize optical disk mount operations, prepare hard media for distribution, issue messages to the (remote) users and archive operator(s). Currently, the above tasks are done manually; the software that will perform them automatically, the Request Handler, is currently being designed.

Acknowledgements

DMF is a project being developed among STScI, ST–ECF and, lately, CADC.

The following people have been, or are currently, involved in DMF development: Rudi Albrecht, Piero Benvenuti, Richard Bowles, Andy Czeko, Daniel Durand, Leslie Hunt, Gerald Justice, Rajesh Kanungo, Steve Lubow, Tom McGlynn, François Ochsenbein, Fabio Pasian, Bill Pence, Benoît Pirenne, Sergio Restaino, Alan Richmond, Fred Romelfanger, Minick Rushton, Guido Russo, Peter Shames, Ethan Schreier, David Silberberg, Claudio Vuerli, Lucy Willard, Andy Woodsworth, Steve Zeller, together with other contributions. DMF operations at STScI are handled by the DSO Branch of the SCARS division.

References

[1] T. McGlynn & L. Hunt (1988), DMF File Handler Guide - Design and Software Specification ST–ECF O-02 Document, Vol. IX.

[2] T. McGlynn, L. Hunt, S. Restaino, F. Romelfanger, G. Russo and L. Willard (1988), Connecting an optical disk archive with a relational database catalogue. In *Online Information 88*, London, U.K.

[3] F. Pasian & F. Romelfanger (1989), The operator's interface to the DMF. ST–ECF technical report *TR(ARC)-10*.

[4] A. Richmond, T. McGlynn, F. Ochsenbein, F. Romelfanger, and G. Russo (1987), The design of a large astronomical database system. In *Astronomy from Large Databases: Scientific Objectives and Methodological Approaches*, ST–ECF/ESO, Garching, Fed. Rep. of Germany.

[5] D. Silberberg, T. McGlynn and S. Lubow (1989), Design of the Data Management Facility Catalog. STScI Technical Document.

[6] D. Silberberg, F. Pasian, B. Pirenne, P. Shames (1989), Design of the DMF Catalog User Interface. STScI Technical Document.

[7] STARCAT User Guide. ST–ECF O-02 Document, Vol. VI, Version 2.25, July 1989.

7

Database Aspects of the Guide Star Catalog

Helmut Jenkner[1]
Space Telescope Science Institute
3700 San Martin Drive
Baltimore, MD 21218, USA

7-1 Introduction

The Hubble Space Telescope (HST) uses off-axis guide stars to achieve its pointing performance. The selection of these guide stars uses a catalog specifically constructed for this task, the Guide Star Catalog (GSC). The first version of the GSC was published in 1989; its production and characteristics are described in detail in Lasker *et al.* (1990, Paper I), Russell *et al.* (1990, Paper II), and Jenkner *et al.* (1990, Paper III).

This article briefly describes the construction of the GSC as necessary to establish the technical foundations, and then discusses the various aspects related to data organization and database structure with regard to the catalog itself and to the required ancillary data.

7-2 History

The Guide Star Catalog has been developed as part of the Guide Star Selection System project, one of the earliest activities of the Space Telescope Science Institute (ST ScI), dating back to its very beginning in 1981. This

---

[1] Affiliated with the Astrophysics Division, Space Science Department, European Space Agency.
Space Telescope Science Institute is operated by the Association of Universities for Research in Astronomy (AURA), Inc., for the National Aeronautics and Space Administration (NASA).

was necessitated by the requirement for early studies and the long lead-times for the photometry program, plate acquisition, and plate scanning. Plate scanning commenced in mid-1984, with a resolution of 50μm. In late 1984, this was changed to 25μm to allow for improved positional and photometric information, as well as for reliable object detection and classification in crowded fields.

Extensive software development was necessary for the required highly automated system (both for catalog construction and the guide star selection functions as part of the HST ground system). Software implementation for the production system started in 1983. Major achievements were the completion of the catalog construction system in June 1984, and the preliminary request processing system in October 1985. The entire software system contains about 200,000 executable source lines (excluding comments).

In June 1989, GSC Version 1.0 has been issued on a set of two digital Compact Discs (CD–ROMs), containing totals of 599 and 590 Mbytes of data, respectively. This version is comprised of 18,819,291 objects with a total of 25,126,027 entries (objects which appear in overlapping survey plate areas may have more than one entry) in the seventh to sixteenth magnitude range; more than 15 million of these objects are classified as stars.

In order to maintain the required positional accuracy over the lifetime of HST, and to allow the identification of faint targets, new northern and southern survey plates will be incorporated over the coming years, leading to a second-generation GSC which will include proper motions and potentially colors.

7–3 Catalog Construction

The GSC is primarily based on an all-sky, single epoch, single passband collection of Schmidt plates. For centers at +6° and north, a 1982 epoch "Quick V" survey was obtained by the Palomar Observatory, while for southern fields, materials from the UK SERC J survey (epoch approximately 1975) and its equatorial extension (epoch approximately 1982) were used. The plates were digitized into 14000 × 14000 rasters at 25μm sample intervals using modified PDS microdensitometers.

The sky-background was approximated with a bi-dimensional cubic spline approximation to the modal level. Then an object finder, based on locating connected pixels at a certain threshold above the background, was used to obtain a list of positions, sizes, intensities, and related descriptive parameters. Images with multiple peaks were deblended by correlations against a library of stellar images.

The identified objects were classified as stars or non-stars by an iterative multivariate Bayesian classifier that used image features from the object-detection steps and was started from a small set of objects visually identified on each plate. Comparison of classifications from multiply cataloged objects in the plate overlap areas shows that the purity of objects classified as stars is typically 97 to 99 percent.

The GSC calibrations are obtained, on a plate-by-plate basis, by polynomial modeling against the photometric and astrometric reference catalogs.

Photometry is available in the natural systems defined by the individual plates in the GSC collection (generally J or V); the calibrations are done using B, V standards from the Guide Star Photometric Catalog (Lasker, Sturch, et al. 1988). The overall quality of the photometry near the standard stars is estimated from the fits and other tests to be 0.15 mag (one sigma, averaged over all plates), while the quality far from the sequences is estimated from the all-sky plate-to-plate agreement and from comparisons with independent photometric surveys to be about 0.30 mag (one sigma), with about 10% of the errors being greater than 0.50 mag.

Astrometry, at equinox J2000, is available at the epochs of the individual plates used in the GSC; the reductions to the reference catalogs (AGK3, SAOC, or CPC, depending on the declination zone) use third order expansions of the modeled plate and telescope effects. The operationally relevant relative astrometric accuracy over a scale length of 30 arcmin. is estimated at 0.3 to 0.4 arcsec. for stars fainter than about 10 mag. Estimates of the overall external astrometric error, produced by comparisons of independently measured positions without regard to location on the GSC plates, are in the range 0.4 to 0.6 arcsec.

7–4 Production Data

Since the GSC construction involved a significant number of individual processing steps for more than 1500 plates, extreme care had been devoted to a flexible but efficient design for the various data items and databases related to individual processing steps, quality assurance, archiving, and the GSC itself.

Control data for the GSC generating processes, the associated raw and final processing data, and selected intermediate data processing steps are contained in the following set of files and databases:

- *Parameter Files* are used to specify nearly all information used to control a program, the occasional exception being information that programs acquire by operator interaction. Parameter files are generically defined for the whole system; as each plate is processed, a plate-specific copy of the system file is made. Production difficulties are generally corrected by modifying the plate-specific parameter file and repeating a computation. Parameter files are formatted in text, so as to facilitate editing and review.

- *Pipeline Files* are those which are used to transfer data from the output of one program to the input of the next within a sequence of computations and also to record important statistical information. Pipeline files, being large quantities of rigidly formatted information, are generally written with fixed-length binary records.

Documentation of the production work is supported by permanently storing the plate-specific parameter files and pipeline files on magnetic disks. Files that contain quality assurance statistics are generally written in text for printing or in metacode for plotting; these relatively voluminous and non-critical files are archived only on magnetic tape.

- *Image Files*, which contain the pixel data from the PDS scans, are a subset of the pipeline files distinguished by their size (typically 400 Mbyte). These are organized as fixed-length binary records with a length of one scan line (typically 14000 pixels, 28 Kbytes), and their contents are described by accompanying header files, which are formatted in the style of FITS headers. Initially, the image data were archived in DEC BACKUP format on magnetic tape; later, these data were copied to optical disks (2 Gbyte each) in strict FITS format, forming the *Image Archive*.

- *Catalog Files* contain the final results of GSC processing. In the ST ScI production version of the GSC, maintenance and enhancement of the catalog requires that these files be stored on-line on magnetic disk in compressed binary form; in the export versions of the catalog on CD–ROM (Compact Disk, Read-Only Memory) or on equivalent magnetic media the catalog files are formatted as standard FITS tables.

In addition to these files, which are used in processing the individual GSC plates, four databases are used to manage the use of the plate collection in catalog generation and to support catalog maintenance and publication. These databases, which are organized primarily by the GSC field identification number, are as follows:

- The *Control Database* is filled with processing status information as the sequence of GSC processing steps is applied. Additionally, it indexes the six-thousand volume magnetic tape collection used in GSC production.

- The *Operator Database* is a manually maintained master list of information related to the production management of the GSC resources. Part of these data, specifically in the area of processing status, are redun-

dant with that in the control database; but additionally, this database supports free-form annotations that the operators use in tracking complex conditions. It is also the repository of information on plates and catalog data, which, while not used in the final GSC, are to be retained for problem solving or quality control.
- The *Recalibration Database* contains the revised calibration information which the generalized catalog server uses to modify the photometric or astrometric reductions as recalibrations are applied. The export version of the GSC was prepared using this feature.
- The *Publication Database* contains the information that describes the statistical and structural properties of the GSC as required for catalog baselining and publication. The publication database is prepared by utilities that combine information from (1) the control database, (2) the operator database, (3) the operators' notebooks, (4) the programs that assembled GSC statistics after the completion of a given version of the catalog, and (5) the recalibration database. For publication, the data contained in the publication database are converted into a set of FITS tables.

7–5 GSC Structure and Content

Access to the GSC has to be accommodated by identification as well as by celestial area. Together with the given large volume of data, this led to the formulation of a structure which would allow fast access, yet constrained by limited storage space and specific software components.

The GSC is organized into 9537 regions; each region therefore holds a few thousand individual object entries. In order to account for the varying population of the catalog as a function of galactic latitude, the region size is not fixed, but varies, so that the number of objects per region is nearly constant. In the on-line versions of the GSC, the region files are implemented as direct-access files with fixed-length records.

Each object in the GSC carries a 10-digit number as identification. The first five digits encode the region number; the last five specify the number of the star within the particular region. Since only a few thousand objects are contained in each region (leaving some room for future expansion), this identification scheme is noncontiguous, *i.e.* there are many 10-digit numbers which do not correspond to an entry in the catalog.

Although this nomenclature as such does not encode coordinates, the region structure of the GSC would provide a hidden coordinate-encoding if the region boundaries were fixed; assuming coordinate updates, this would lead to the well-known problems of coordinate-related nomenclature techniques. Therefore, an additional feature has been adopted for the GSC regions: they are not (primarily) defined by their celestial coordinates, but by their content. In other words, once an object has become a member of a region, it remains part of it regardless of position updates. Therefore, the maximum and minimum coordinates of regions may overlap slightly, once updating activity has occurred.

Each region is thus characterized by its center (reference) position and extent, as well as by the number of objects it contains. (Back-pointers to the individual plates from which objects appear in the region as well as several organizational items are available in addition.) For each entry the following information is contained in the GSC:

- Identification (GSC Number).
- Position, in the form of offsets (ξ, η) to the region reference position.
- Position standard deviation (*one* per position).
- Magnitude band descriptor.
- Magnitude and magnitude standard deviation.
- Object classification (star versus non-star).
- Source plate (or catalog) identification.
- Various flags.

Maintenance of catalog identifications is facilitated by associating two flags with each object. The *published* flag signifies that, as part of the GSC distribution process, an object has been frozen into the catalog and its name reserved. Published objects are never physically deleted from the on-line reference version of the catalog; rather, when it is necessary to remove one of them, its *deleted* flag is set, signifying that the records associated with the entry are not astronomically part of the GSC and that the name is reserved against future use. An example of a published and deleted object is an image-processing defect identified after the catalog was frozen. Published but deleted objects are generally of interest only for catalog maintenance; except when special switches are set, they are not read by GSC utilities in the HST ground system or in maintenance and support activities.

Note also that object data derived from different plates (*i.e.* for objects appearing on more than one plate) are stored separately, but with the same identification.

7–6 Enhancements and Plans

Over the next years, improved versions of the GSC will be derived and issued, containing improved calibrations, both with regard to astrometry and photometry, the correction of known defects and artefacts, and cross-references to existing catalogs. While these maintenance activities will not significantly alter the scope of the GSC as it exists today, the incorporation of new surveys is also planned. Given these new northern and southern second-epoch surveys, it will be possible to derive proper motions as well as colors, if plate passbands different from the first version are selected. For this second-generation Guide Star Catalog, preparations to significantly improve the image processing techniques have begun, based on the experiences obtained in generating the first version of the catalog.

From the organizational (database) point of view, we anticipate an enhanced catalog structure for the future, which will, however, retain the basic GSC nomenclature and region scheme. This new structure is based on the need to:

- preserve raw data (image features) obtained directly from the underlying plates;
- perform re-calibration of selected plates or celestial areas without affecting the rest of the catalog;
- provide corrections in an orderly and trackable way; and
- provide cross-references to other catalogs.

To achieve this goal, it is expected that the main part of the catalog will contain object names, image features, and other information, directly derived from the plates, but without any additional calibration. A second part will contain these constructs, *e.g.* plate solutions and photometric transformations. Addenda and error corrections will be carried in a third database. Cross-references to other catalogs (as well as the catalog data themselves) will also be part of this new structure. The various pieces of information related to one object would be combined by a software module (the catalog server) upon demand, *i.e.* upon a request by a user for data from the catalog. This enhancement would allow the efficient incorporation of new survey data, and also of other catalog data as they become available in the community.

Acknowledgements

Due to the very nature and size of the Guide Star Selection System project, which contained —as an integral part— all the efforts leading to the generation of the Guide Star Catalog, a large team of scientists, software professionals, engineers, operators, and (last but not least) managers, was required to cover all the different areas of expertise needed. Therefore, this paper reflects the efforts of all the past and present staff members of the Space Telescope Science Institute who have been and still are involved in this endeavor, as well as the contributions from a number of distinguished colleagues in the international astronomical community; since it would be impossible to mention all contributors in this context, the reader is referred to the corresponding section in Paper III.

A complete list of publications related to the Guide Star Catalog can be found in Papers I through III, as well as in Jenkner et al. (1988).

References

[1] Jenkner H., Russell J. L., Lasker B. M. (1988), The Guide Star Catalog III. Structure and Publication, Status and Plans; in: *"Mapping the Sky – Past Heritage and Future Directions"*, IAU Symp. **133**, S. Debarbat, J. A. Eddy, H. K. Eichhorn, & A. R. Upgren (Eds.), 239.

[2] Jenkner H., Lasker B. M., Sturch C. R., McLean B. J., Shara M. M., Russell J. L. (1990), *Astron. Journal*, **99**, No. 6, 2082 [Paper III].

[3] Lasker B. M., Sturch C. R., Lopez C., Mallama A. D., McLaughlin S. F., Russell J. L., Wiśniewski W. Z., Gillespie B. A., Jenkner H., Siciliano E. D., Kenny D., Baumert J. H., Goldberg A. M., Henry G. W., Kemper E., Siegel M. J. (1988), *Ap. J. Suppl.* **68**, No. 1, 1.

[4] Lasker B. M., Sturch C. R., McLean B. J., Russell J. L., Jenkner H., Shara M. M. (1990), *Astron. Journal* **99**, No. 6, 2019 [Paper I].

[5] Russell J. L., Lasker B. M., McLean B. J., Sturch C. R., Jenkner H. (1990), *Astron. Journal* **99**, No. 6, 2059 [Paper II].

8

The HIPPARCOS INCA Database

C. Turon, F. Arenou, M.-O. Baylac, D. Boumghar, F. Crifo,
A. Gómez, M. Marouard, M. Mekkas, D. Morin, & A. Sellier
Bâtiment Hipparque
URA 335 et GDR 51 du CNRS
Observatoire de Paris
92195 Meudon Principal Cedex, France

The Hipparcos INCA database was created at the initiative of D. Morin and has been developed by F. Arenou and D. Morin, with the help of M.-O. Baylac, M. Marouard, M. Mekkas, and A. Sellier, under the scientific supervision of A. Gómez, F. Crifo and C. Turon. The reconversion of the database from the Orsay Computer center to the Ultrix DEC system presently used was done in collaboration with the SIMBAD team at Strasbourg, mainly by F. Arenou and D. Boumghar.

This work was performed in close cooperation with the CDS staff, and the contributions of D. Egret, F. Ochsenbein, and M. Wenger should be especially acknowledged.

8–1 The HIPPARCOS mission: requirement for an INCA Database

The HIPPARCOS satellite, dedicated to the measurements (to the milliarcsecond level) of positions, trigonometric parallaxes, and proper motions for about 120,000 stars, was included in the scientific programme of the European Space Agency in 1980. It was successfully launched by Ariane on 8 August 1989, but could not reach the geostationary orbit due to the failure of its apogee boost motor. The satellite was placed on a highly elliptical orbit, very close to the geostationary transfer orbit (perigee about 500 km, apogee, 36 000 km, period, $10^h \, 40^{min}$). A *revised mission* was defined, which kept, as much as possible, the characteristics of the nominal mission. In particular, it was decided to keep the observing programme, the HIPPARCOS *Input Catalogue*, unchanged.

Due to the operating mode and to the detection system of the satellite, the observing programme had to be constructed in advance. Moreover, to allow a precise pointing of the detector (an image dissector tube) and a correct estimation of the observing time to be allocated to each star of the programme, an *a priori* knowledge of positions and magnitudes was required to a precision of 1.5 arcsecond and 0.5 magnitude respectively. As a consequence, a systematic collection and analysis of all available data on all stars proposed for observation was undertaken, and extensive campaigns of preliminary ground-based observations of positions, magnitudes and colours were performed before the satellite launch (Turon, 1988a). The creation of the INCA database, from the SIMBAD database (see chapter 9), was the clue for an efficient handling of all these data.

The *Input Catalogue* was constructed from the database, taking benefit of its archiving and analysis capabilities. It contains the best data available in the database for each considered quantity. The selection of target stars was optimized through successive iterations, using simulations of the observation by the satellite. The resulting catalogue exploits at best the observing capabilities of the satellite, while satisfying as much as possible a very large variety of scientific programmes.

For a detailed description of the satellite and of the Input Catalogue, see *The Hipparcos Mission, Pre-launch Status*, **I** and **II**; the INCA Colloquia **1**, Aussois (1985), and **2**, Sitges (1988); and Turon (1987), Turon (1989a).

1.1 Proposals for observation with HIPPARCOS

In answer to an *Invitation for proposals*, 220 observing programmes, dealing with a large variety of astronomical topics, were submitted from the world–wide astronomical community to ESA by the end of 1982. A Programme Selection Committee appointed by ESA rated these proposals with respect to their scientific interest and their adequacy to HIPPARCOS capabilities. The lists of stars proposed within the frame of these proposals were transmitted to the INCA[1] team in the Observatoire de Paris–Meudon, and the first work, besides the reading of this huge amount of cards and tapes (about 750,000 entries were submitted, each star being possibly requested by up to 20 different identifiers !), was to obtain a unique list of stars. This work was first, and mainly, performed by an intensive use of the SIMBAD database (operated at this time on the UNIVAC computer of the Strasbourg-Cronenbourg Computer Center), and thanks to the efficiency of its query software. For each star, all the cross-identifications found in SIMBAD were kept, along with the references of each proposal requiring

---

[1] INCA is the Consortium of astronomical Institutes which was charged by the European Space Agency to elaborate the HIPPARCOS Input Catalogue. Team Leader: C.Turon.

the observation of this star (Gómez & Crifo, 1985; Turon, Gómez, Crifo & Grenon, 1989).

It rapidly appeared that the best way to handle and update this large amount of data (the data coming from SIMBAD, but also all the data to come from new compilations and observations) was to use an efficient database software, which would allow to keep the memory and reference of all the data, to have a rapid access to any star by any identification, to make any useful selection among the data, and any sorting, testing, updating, displaying, *etc*. It was therefore decided, by the end of 1984, to use a possibility foreseen by the SIMBAD software, *i.e.* to create a sub-base of SIMBAD which would have the possibility to evolve by its own, and also to be considered as a 'private' working tool, available only, at least until it is ready for publication, to the INCA team.

8–2 The INCA database: creation

The INCA database was therefore created in January 1985 (Morin & Arenou, 1985; Arenou & Morin, 1988; Gómez, 1988; Turon, Gómez, Crifo & Grenon, 1989; Gómez, Morin & Arenou, 1989), on the UNIVAC computer of Paris-Sud-Informatique (Université Paris XI, Orsay) where SIMBAD was operated at the time. At its creation, the INCA database included all the stars proposed for HIPPARCOS observation and identified in SIMBAD, with all the data available in SIMBAD for these stars, with exception of the bibliography: cross-identifications, *basic data*, and *measurements*. The stars not found in SIMBAD were added to the new database, using specific INCA identifiers, and, when available, identifiers given by the proposers.

A specific software was developed by the INCA team in order to search within SIMBAD for stars proposed for HIPPARCOS observation under names not included in SIMBAD. Possible cross-identifications were first manually examined and, when necessary, submitted to the proposers for approval. About 40,000 stars were processed in this way. Finally, of the 200,000 proposed stars, only 4,000 stars were not recognized in SIMBAD.

In addition to these stars coming from one or another of the 220 proposed observing programmes, a basic list of bright stars ("survey") required for satellite operation and data reduction was also included in the INCA database. The stars of this "survey" were selected from SIMBAD according to the following limiting magnitudes:

$V_{lim} = 7.9 + 1.1 \times |sinb|$ for spectral types earlier than or equal to G5;
$V_{lim} = 7.3 + 1.1 \times |sinb|$ for spectral types later than G5.

The content of the survey was updated later, as new photometry became

Spectral Types	V magnitudes				Total	%
	< 6	6 – 8	8 – 10	≥ 10*		
O - B	1100	4700	13600	6200	25600	12
A0 - A9	1000	8000	21400	4400	34800	16
F0 - F9	700	6900	30400	10000	48000	22
G0 - K1	1200	12100	40800	10000	64100	30
K2 - M8	1100	6500	17600	4600	29800	14
others					12400	6
Total	5100	38200	123800	35200	214700	100

* no star fainter than $V = 12.5$

Table 8–1: Magnitude and spectral type distribution of INCA stars

available (Crifo et al., 1985; Turon, Gómez, Crifo & Grenon, 1989; Gómez, Crifo & Turon, 1989).

8–3 The INCA database: stellar content

With the exception of the survey whose stars were included in a systematic way with the objective of being complete up to a given magnitude, the stellar content of the INCA database is the result of the addition of lists of stars proposed for very different scientific reasons (Turon et al., 1988; Turon, Gómez, Crifo & Grenon, 1989; Gómez, Crifo & Turon, 1989, for stars retained in the HIPPARCOS observing programme).

The only common rationale in the choice of these stars is that they are supposed to have a trigonometric parallax or a proper motion measurable at the milliarcsecond level (with the exception of the stars proposed for reference systems).

3.1 Magnitude and spectral type distribution.

The magnitude and spectral type distribution of the nearly 215,000 stars of the INCA database is given in Table 8–1.

Many categories of peculiar stars can be found in the INCA database, as a consequence of the variety of the astronomical programmes proposed for observation with Hipparcos. Table 8–2 shows their distribution.

(a) Peculiar stars nearer than 400pc	
Be	1200
Ap, Am	4800
Other peculiar stars	2700
(b) Variable stars	
Cepheids	340
RR - Lyrae	370
Long period	1370
Irregular	490
(c) Miscellaneous	
Wolf-Rayet	80
High latitude O-B stars	890
Central stars of planetary nebulae	50
White dwarfs	40
X-ray binaries	30
(d) Double and Multiple stars	
Double systems	23660
Multiple systems	5980
Visual binaries	1200
Eclipsing binaries	100
(e) Galactic clusters	
In 280 galactic clusters	4000
(f) Magellanic Clouds	
Small Magellanic Cloud	90
Large Magellanic Cloud	370

Table 8–2: Various categories of stars included in the INCA database

3.2 Double and multiple systems: a special processing

A specificity of the INCA database with respect to SIMBAD is the processing of double and multiple systems. This processing had to take into account, for all possible configurations of double or multiple systems, the point to be observed by the satellite. Due to the shape of the "Instantaneous Field of View", the observation of a star may be perturbed by the presence of a neighbouring companion, as a function of the distance and difference of magnitude between them (Turon, Gómez, Crifo & Grenon, 1989; Turon, Kovalevsky & Lindegren, 1989).

For distances smaller than 10 arcseconds, a unique target is given to the satellite. A unique entry is therefore appearing in the INCA database, which may be the primary component, the photocentre or the geometric centre of the system depending on the configuration of the system. For larger separations, multiple entries may have been kept (possibly after the

Category	Double Systems	Multiple Systems
One entry ($\rho \leq 10$ arcsec)		
primary component	7050	1290
geometric centre	1100	* 230
photocentre	4140	830
Total	12290	2350
Two or more entries ($\rho > 10$ arcsec)	11370	3630
Total	23660	5980

* In case the third (or further) components can be neglected.

Table 8–3: Different categories of double and multiple systems in the INCA database

inclusion of perturbing components in the INCA database even though they were not requested by any scientific programme), and a specific observing mode may have been flagged in some cases (alternate pointing of two components perturbing each other). The distribution of double and multiple systems in these different categories is shown in Table 8–3.

3.3 Some catalogues totally or partially included.

A number of catalogues were proposed as a whole (mainly astrometric catalogues), with the aim of improving the materialization of a reference system, or to link existing systems with the future HIPPARCOS system: FK5 (Fricke *et al.*, 1989), and FK5 bright extension, N30 (Morgan, 1952), NPZT (Yasuda *et al.*, 1982), IRS (AGK3R (Corbin, 1978) + SRS (Polozhentsev, 1978; Hughes, 1978) + supplement), AGK3 stars already part of the Lick Proper Motion Survey (Klemola *et al.*, 1987), GC (Boss, 1937).

Other catalogues which are included to a high percentage are the catalogues of nearby and high proper motion stars: Gliese's catalogue of stars nearer than 22 parsecs (Gliese, 1969) and extensions (Gliese and Jahreiss, 1979), and LHS catalogue (stars with proper motions larger than 0.5 arcsecond per year, Luyten, 1976), are completely included up to the HIPPARCOS limiting magnitude; a high proportion of NLTT (stars with proper motions larger than 0.18 arcsecond per year, Luyten, 1980) and of Mc Cormick red dwarf stars are included.

Some catalogues established for astrophysical purposes are also highly represented in the INCA database: Catalogue of extinction data (Neckel *et al.*, 1980), Catalogue of [Fe/H] measurements (G. Cayrel *et al.*, 1982), Catalogues of Central stars of planetary nebulae (Perek and Kohoutek, 1967; Acker and Gleizes, 1982), the Michigan Spectral Survey (Houk & Cowley, 1975; Houk, 1978, 1982), *etc.*

8–4 The INCA database: data content

4.1 The identifiers

In addition to all the identifiers coming from SIMBAD at the time of creation of the INCA database, new identifiers, proper to INCA, were introduced (Gómez, 1988; Gómez, Morin & Arenou, 1989):

- identifiers giving, for each of the proposed observing programme, the running number in the proposal, and the priorities allocated to this star by the proposer him(her)self, and also, by the ESA Programme Selection Committee;
- the CCDM numbers, referring to the "Catalogue des Composantes d'étoiles Doubles et Multiples" (Dommanget, 1983);
- the LID numbers referring to the identifiers used at the Institut d'Astronomie de Lausanne for their database of photoelectric photometry (Mermilliod & Mermilliod, 1982; Mermilliod, 1984).

Finally, during the whole process of the construction of the Input Catalogue, a very large number of checks and corrections were performed on the various cross-identifications. Mainly, tests of the coherence between data and identifiers were systematically achieved (Gómez, 1988; Gómez, Morin & Arenou, 1989): HD numbers should increase with α_{1900}, DM numbers should increase with α_{1875} or α_{1855} in a given one-degree zone, *etc.*; if a star has a LHS identifier, its proper motion should be higher than half a second per year; if a star has a HR identifier, its V magnitude should be brighter than 6.5 (except if it is a nova !), *etc.* The identifiers within multiple systems, and the cross-identifications within galactic clusters were also systematically checked (and had to be fixed often !).

4.2 The data

At the creation of the INCA database, all data available in SIMBAD for all the stars common to both bases were incorporated into INCA. Later on, all the data compiled or newly observed within the frame of the work of the INCA Consortium were regularly introduced in the database, after systematic cross-checks with the existing data (Gómez, 1988; Gómez, Morin & Arenou, 1989):

- compilation of positions and proper motions, hierarchisation of the data (Jahreiss, 1988; Jahreiss, 1989),
- new observations of positions with Automatic Meridian Circles, and new measurements on sky survey or astrograph plates (Réquième, 1988a; Réquième, 1988b; Réquième, 1989; Muiños et al., 1988; Tucholke, 1988),
- compilation of magnitudes and colours (Mermilliod & Mermilliod, 1985),
- new observations of magnitudes and colours (Grenon, 1988a; Grenon, 1988b; Grenon, 1989),
- estimation of colours from spectral types and an improved model of interstellar extinction (Gómez, Morin & Arenou, 1989),
- compilation and new observations of the parameters of double and multiple systems: separation between components, magnitudes of components, variability (Dommanget, 1988; Dommanget, 1989),
- computation of mean magnitudes for variable stars, as a function of the variability type (Mennessier & Baglin, 1988; Mattei, 1988; Mennessier & Figueras, 1989).

In addition, all the results of the successive numerical simulations of the observation were also included in the database.

In order to minimize any possible waste of observing time with HIPPARCOS, a great care was devoted, especially for the programme stars, to guarantee the quality and reliability of the data. Extensive tests were systematically performed to cross-check the consistency of all data (acceptable extreme values, consistency between different data for the same kind of stars, consistency of the data with the considered scientific programme, *etc.*).

8-5 Usage of the INCA database

During the preparation of the Input Catalogue, an extensive use of the INCA database has been made by the ten persons of the INCA team at Meudon Observatory: about 55,000 star entries read per day and 4 600 modified per day (or, which is equivalent, each star entry has been read 380 times and updated 32 times in four years).

Once edited, the INCA database contains 420 Mbytes of data, among which 2,492,000 identifiers and 6,561,000 measurements for the about 215,000 stars. Although INCA was created as a subset of SIMBAD, it has evolved so much that only 24% of identifiers and measurements are now coming from SIMBAD, the other 76% being data newly compiled, obtained, or computed by the INCA consortium.

8–6 Database software and organization of the data

Initially the INCA database, as well as SIMBAD, were hosted on the UNIVAC computer at Orsay campus. The software was written in PL/1 and in assembler code. The UNIVAC had to be stopped, and a conversion (rewriting in fact) of the database software was then necessary for both databases (see chapter on SIMBAD, this book). As INCA was created as a subset of SIMBAD, it could also take profit of the new software. In consequence, the INCA team has collaborated to the rewriting; in particular, CNES provided a 2-year work contract for an engineer whose work was especially dedicated to this task.

The rewriting of the database code required the choice of an adequate operating system and of a widely used language. One of the main constraints was the portability of the software between different machines. The choice made eventually by the CDS (see chapter 9) was to use UNIX and C. The concept of Object Oriented Programming (OOP) seemed to be the best one to ensure good coordination and integration of the programmers' work and ease later debugging and extension.

The data in the INCA database are organized as those of SIMBAD: INCA contains identifiers, basic data and measurements, like SIMBAD, but not the bibliography. Apart from the role of updating the database and providing the best data to ESA for Hipparcos operations, our work also deals with the analysis of the database contents; current software developments are related with the visualization of data and statistics.

8–7 Access mode

The INCA database is now implemented at Meudon Observatory on a file server DS5400 of Digital Equipment Corporation under Ultrix operating system. This computer is connected to SPAN (node name: MEHIPA, or 17763). However, due to the role of working tool of the INCA database, its access is —up to now— restricted to INCA consortium members.

8–8 Future plans.

The organisation of the HIPPARCOS INCA data in database greatly eases the long and iterative preparation work for the Input Catalogue. It allows, all along the work, to take into account very rapidly the newly compiled or observed data, to make very efficiently all necessary sortings and tests, to make displays of the all-sky distribution of any parameters, to display the choice of stars in small areas of the sky and help understanding the possible problems and necessary modifications, to check the configurations of multiple systems or galactic clusters obtained with the positions included in the database, and compare them with the available charts, *etc.*

The first aim was the construction of the HIPPARCOS Input Catalogue, and the first publication will be the Input Catalogue itself, where the best data in each category were carefully chosen (Turon, 1988b; Perryman & Turon, 1988; Turon, 1989b). Its publication is scheduled for early 1991. The catalogue will be made available on paper (through ESA distribution), on tape (through CDS distribution), and on CD-ROM.

The next step is to make available to the astronomical community the work performed on the stars of the INCA database which are not retained for HIPPARCOS observation (about 100,000 stars). Several solutions are here foreseeable, and the choice is still to be made: the database could be made available as a sub-base of SIMBAD, or the data could be incorporated in SIMBAD as "INCA measurements or compilations".

References

[1] Acker A. & Gleizes F. (1982), *Special CDS Publication*, 3.
[2] Arenou F. & Morin D. (1988), ESO Conf. Workshop: *Astronomy from large databases*, Garching, 12-14 oct. 1987, F. Murtagh & A. Heck (Eds.), 269.
[3] Boss B. (1937), Carnegie Institution of Washington Pub.
[4] Cayrel de Strobel G., Bentolila C. & Hauck B. (1982), Magnetic tape from CDS.
[5] Corbin T.E. (1978) IAU Coll. 48, 505.
[6] Crifo F., Turon C. & Grenon M. (1985), INCA Colloquium 1, 67.
[7] Dommanget J. (1983), *Inform. Bull. CDS* 24, 83.
[8] Dommanget J. (1988), INCA Colloquium 2, 191.
[9] Dommanget J. (1989), ESA SP-1111, II, 149.
[10] Fricke W. *et al.* (1989), *Veroff. Astron. Rechen Institut Heidelberg*, 32.
[11] Gliese W. (1969), *Veroff. Astron. Rechen Institut Heidelberg*, 22.
[12] Gliese W. & Jahreiss H. (1979), *Astron. Astrophys. Suppl.*, 38, 423.
[13] Gómez A. & Crifo F. (1985), INCA Colloquium 1, 57.
[14] Gómez A. (1988), INCA Colloquium 2, 33.
[15] Gómez A., Crifo F. & Turon C. (1989), ESA SP-1111, II, 89.
[16] Gómez A., Morin D. & Arenou F. (1989), ESA SP-1111, II, 23.
[17] Grenon M. (1988a), *International Conference on Coordination of observational projects in astronomy*, Strasbourg 23-26 nov. 1987, C. Jaschek & C. Sterken (Eds.), Cambridge Univ. Press, 65.
[18] Grenon M. (1988b), INCA Colloquium 2, 343.
[19] Grenon M. (1989), ESA SP-1111, II, 129.
[20] *The Hipparcos Mission, Pre-launch Status*, I, ESA SP-1111 (1989), M.A.C. Per-

ryman & H. Hassan (Eds.).
[21] *The Hipparcos Mission, Pre-launch Status*, II, ESA SP-1111 (1989), M.A.C. Perryman & C. Turon (Eds.).
[22] Houk N. & Cowley A. P. (1975), Catalogue of two dimensional spectral types for the HD stars, 1, Univ. Michigan, Ann Arbor.
[23] Houk N. (1978), *ibid.*, 2.
[24] Houk N. (1982), *ibid.*, 3.
[25] Hughes J.A. (1978), IAU Coll. **48**, 497.
[26] INCA Colloquium 1, Aussois, France, 3-7 juin 1985, ESA SP-234, C. Turon & M.A.C. Perryman (Eds.).
[27] INCA Colloquium 2, Sitges, Spain, 25-29 janvier 1988, C. Turon & J. Torra (Eds.).
[28] Jahreiss H. (1988), INCA Colloquium 2, 289.
[29] Jahreiss H. (1989), ESA SP-1111, II, 115.
[30] Klemola A. R., Jones B. F. & Hanson R. B. (1987), *Astron. J.*, **94**, 501.
[31] Luyten W. J. (1976), LHS Catalogue, Minnesota.
[32] Luyten W. J. (1980), NLTT Catalogue, Minnesota.
[33] Mattei J. (1988), INCA Colloquium 2, 379.
[34] Mennessier M.O. & Baglin A. (1988), INCA Colloquium 2, 361.
[35] Mennessier M.O. & Figueras F. (1989), ESA SP-1111, II, 163.
[36] Mermilliod J.C. & Mermilliod M. (1982), in *The scientific aspects of the Hipparcos Space Astrometry Mission*, ESA SP-177, 139.
[37] Mermilliod J.C. (1984), *Inform. Bull. CDS* **26**, 3.
[38] Mermilliod J.C., & Mermilliod M. (1985), INCA Colloquium 1, 111.
[39] Morgan H. R. (1952), *Astron. Papers Amer. Eph.*, **13**, 3.
[40] Morin D. & Arenou F. (1985), INCA Colloquium 1, 63.
[41] Muiños J.L.,Quijano L., Morrison L., Gibbs P., Helmer L., & Fabricius C. (1988), INCA Colloquium 2, 273.
[42] Neckel T., Klare G. & Sarcander M. (1980), *Bull. Inf. CDS*, **19**, 61.
[43] Perek L. & Kohoutek L. (1967), Publ. Czechoslovak Academy of Sciences.
[44] Perryman M.A.C. & Turon C. (1988), INCA Colloquium 2, 409.
[45] Polozhentsev D.D. (1978), IAU Coll. **48**, 489.
[46] Réquième Y. (1988a), *International Conference on Coordination of observational projects in astronomy*, Strasbourg 23-26 nov. 1987, C. Jaschek & C. Sterken (Eds.), Cambridge Univ. Press, 59.
[47] Réquième Y. (1988b), INCA Colloquium 2, 267.
[48] Réquième Y. (1989), ESA SP-1111, II, 107.
[49] Tucholke H.J. (1988), INCA Colloquium 2, 285.
[50] Turon C. (1987), IAU Symp. 133, Paris, 1-5 juin 1987, S. Debarbat *et al.* (Eds.), Kluwer Acad. Press, 245.
[51] Turon C. (1988a), *International Conference on Coordination of observational projects in astronomy*, Strasbourg, 23-26 nov. 1987, C. Jaschek & C. Sterken (Eds.), Cambridge Univ. Press, 47.
[52] Turon C. (1988b), INCA Colloquium 2, 397.
[53] Turon C. (1989a), in *Star Catalogues: A Centennial Tribute to A. N. Vyssotsky*, IAU Comm. 24, Baltimore, 5 August 1988, A.G. Davis Philip & A.R. Upgren

(Eds.), L. Davis Press, 65.
[54] Turon C. (1989b), ESA SP-1111, **II**, 225.
[55] Turon, Gómez A. & Crifo F. (1988), ESO Conf. Workshop: *Astronomy from large databases*, Garching, 12-14 oct. 1987, F. Murtagh & A. Heck (Eds.), 73.
[56] Turon C., Gómez A., Crifo F. & Grenon M. (1989), ESA SP-1111, **II**, 7.
[57] Turon, Kovalevsky J. & Lindegren L. (1989), ESA SP-1111, **II**, 65.
[58] Yasuda H. *et al.* (1982), *Ann. Tokyo Astr. Obs.*, **18**, 367.

9

The SIMBAD astronomical database

Daniel Egret, Marc Wenger, & Pascal Dubois
CDS, Observatoire astronomique
11, rue de l'Université
67000 Strasbourg, France

9-1 Introduction

SIMBAD is the astronomical data base produced and maintained by the Centre de Données astronomiques de Strasbourg (CDS), at the Observatoire de Strasbourg, France.

The acronym SIMBAD stands for Set of Identifications, Measurements, and Bibliography for Astronomical Data. We describe here the present status of the database, the features related to access, usage and updating, and finally describe the expected future developments.

9-2 The Strasbourg Data Center (CDS)

The CDS is located at the Observatoire Astronomique de Strasbourg (France). CDS personnel created and implemented the SIMBAD data bank and maintain its data and software system. CDS is operated under an agreement between French Institut National des Sciences de l'Univers (INSU) and Université Louis Pasteur, Strasbourg (ULP). CDS also centralizes and distributes astronomical catalogues and archives at an international level.

CDS takes an active part in a number of space projects which rely upon SIMBAD for the preparation of their respective input catalogues. As an example, the INCA database of the HIPPARCOS Input Catalogue consortium is organized as a subset of SIMBAD, sharing the same computer resources (see chapter 8, this book).

Information about CDS services and the latest developments can be found in the six-monthly CDS Information Bulletin. The Bulletin also contains general papers about astronomical data, statistical methodology, other data centres' activities, etc. CDS also produces directories gathering all practical data available on associations, societies, scientific committees, agencies, companies, institutions, observatories, *etc.*, more generally organizations, involved in astronomy and/or space sciences [9] — see chapter 22, this book.

9–3 The astronomical contents

SIMBAD is a unique set of basic astronomical data, providing identifications, measurements, and bibliographical references for astronomical objects. Originally created for stellar data, SIMBAD now includes galaxies and non-stellar objects. The only astronomical objects specifically excluded from SIMBAD are Solar System bodies.

SIMBAD was first created by merging the Catalog of Stellar Identifications (CSI) [11] —a dictionary of identifications of stars obtained by cross-matching and merging a number of astronomical catalogues— and the Bibliographic Star Index (BSI) [12] —a survey of bibliographical references from the astronomical literature— as they existed at the Meudon Computer Centre until 1979. The resulting database was then expanded by the addition of source data from the many catalogs connected to the CSI and by new literature references. The first on-line interactive version of SIMBAD was released in 1981, and the database was extended to galaxies and other non-stellar objects in 1983.

SIMBAD now contains: 740,000 entries, including 640,000 stars, and 100,000 non-stellar objects (mostly galaxies). 2,500,000 identifications and more than 1,200,000 measurements are provided on line.

The bibliographic index contains references to stars from 1950 onwards, and to galaxies and all other objects outside the solar system from 1983 onwards. Presently there are about 1,000,000 references taken from 60,000 papers published in the 90 most important astronomical periodical publications.

For each object, the following data are included:

- basic data:
    - stars: coordinates (different systems may be displayed), spectral type, blue and visual magnitudes, and proper motions (including rms errors).
    - galaxies: coordinates, blue and visual integrated magnitudes, morphological type, and physical dimension.
- cross–identifications (or *aliases*) from more than 400 source catalogues

completely or partially included in the data bank; a list of the catalogues used is available on request [13].
- observational data for some twenty different types of measurements (*e.g.* astrometric data, stellar parallaxes, radial velocities, magnitudes or fluxes at different wavelengths, spectral classifications, *etc.*). For each type, one can retrieve: individual data with their bibliographical references and weighted means computed from existing observed values by specialists in each data type.
- general bibliography for the object, including references to all published papers from the journals regularly scanned (currently about 90 titles). The bibliography is reasonably complete back to 1950 for stars, but only to 1983 for non-stellar objects (although many earlier papers are included for certain objects). Articles are scanned in their entirety, and references to all objects mentioned are included in the bibliography.

More details about the way SIMBAD was constructed and about the data sources can be found in Egret, 1986.

9-4 Retrieving data: three query modes

The data, measurements, and bibliography in the SIMBAD database can be retrieved:

- through any object designation (about 400 different types)
- by object coordinates (at any equinox). In this latter mode, one may request to get all objects within a circle of given dimensions around the given position.
- through *filters*, in a sampling mode, where criteria can be specified on parameters such as position, magnitude, existence of various types of data, etc.

More details about the query modes are to be found in the SIMBAD user's guide [17].

Output of data is driven by formats: data presentation is tuned through formatting instructions, on screen and files. Files can be mailed to the user. The user may write his own formats or modify existing ones using a text editor. Such formats can be stored and reused in a later session.

The user interface is adaptable to user preferences. Line by line on simple terminals or with full screen management on more sophisticated terminals. Evolution towards graphical terminals is planned. This user interface runs, for its full screen part, the TERMWINDOWS/PROTEUS ([5], [6]) software packages developed at ESO, Garching.

9–5 SIMBAD III: a new Database Management System

Since the time SIMBAD was first designed, at the end of the 70's, many things have evolved concerning astronomical data base content and design: data volumes have increased, a wide range of computers are now accessible with increased power and lower prices, networking has become common and fast.

These considerations, added to the necessity of leaving the Univac host computer in Orsay which would stop operations in 1990, drove us to redesign SIMBAD, taking into account all these trends (Wenger, 1987; Egret & Wenger, 1988). The new version of SIMBAD (SIMBAD III) was released on July, 1st, 1990. The software has been designed by the SIMBAD team in Strasbourg (Marc Wenger, François Bonnarel, Soizick Lesteven), with the collaboration of the Hipparcos INCA team in Meudon (Frédéric Arenou, Djamel Boumghar) and the kind contribution of François Ochsenbein (ESO). We will give here an idea of the concepts used while producing this modern new software.

5.1 Portability: a primary requirement

Designing a large software system today requires that special attention be given to portability —a very important requirement for ensuring the utility of the enormous amount of work needed for writing software. Such software can commonly be in use ten years and even more: computer technology evolves so rapidly that it will have to run on different hardware during this period.

In order to achieve portability, the new SIMBAD III software has been written in C language and does not use machine and compiler dependent features.

To preserve the software investment, it is also necessary to be able to easily upgrade software, by adding new features and adapting the existing ones to new requirements. Such modifications are facilitated by a well structured design, each function being as independent as possible from any other one. Building structured software is achieved in SIMBAD by using the concept of object-oriented programming (Wenger, 1989).

5.2 An object-oriented data base

The information unit in SIMBAD is the astronomical object. Such an object can be described by one identifier and its coordinates. It can also consist of 30 identifiers, 20 measurement types, each type containing from one to fifty measurements, and 600 bibliographical references. Up to now, no commercial database management system is able to fit easily the requirements concerning such variability in the amount of data for different objects. Although it is theoretically possible to structure SIMBAD using the relational database model, it would need to split an object into so many tables and rows that gathering all information for a well studied object would need several hundred input/outputs.

Thus appears the necessity of creating a specific data base management system for SIMBAD, in order to keep its orientation towards astronomical objects and to make its performance efficient.

The solution adopted for SIMBAD uses the concept of object-oriented programming as described for instance in [7].

An OBJECT is an abstraction made of the encapsulation of the data needed to describe its properties, and a set of functions, here called METHODS, devoted to the manipulation of these data. Objects having the same properties are grouped in CLASSES. An object is manipulated through MESSAGE TRANSMISSION. Such a message consists of the receiver object, the method to activate, and eventually a list of parameters.

Object orientation allows simple software design (by specifying each object), leads to modularity (objects interact only through message transmission), ensures data integrity (they are concealed in the object definition) and facilitates interface with other software (objects are manipulated by a single and simple concept: the message transmission).

To maximize performance and portability, object management in SIMBAD has been implemented through a tool kit of standard C language functions.

5.3 Commands, agents and verbs

In order to avoid a possible confusion between *astronomical objects* and *computer objects*, we have adopted for describing SIMBAD a slightly different terminology: we introduced *commands, agents, verbs*, and *parameters*.

A *command* is the complete instruction sent to the programme for execution. Every command has the following syntax:

AGENT VERB PARAMETERS.

The *agent* is the main concept in SIMBAD: all requests from the user will be sent to agents. These requests are made by use of a *verb* and optional parameters.

A command example is: Simbad search SN 1987A. (Agent: Simbad; verb: search; parameter: *the object name*).

Each agent 'knows' how to execute several commands, depending on its abilities. The same verb can also be used for different agents: `display`, for instance, can be used to show the instructions in a format or a filter, or to list the data available for an astronomical object.

Agents having the same abilities are grouped in *categories* (classes). Examples of categories are: ASTRONOMICAL OBJECT, OBJECT LIST, DATABASE. Several agents of a same category may exist in a session. For example, we are able to use several formats simultaneously, compare different astronomical objects and even query more than one database in the same session.

Complete SIMBAD software capability is contained in the 'knowledge' of the available agents, the action verbs for each agent and the possible parameters for each verb. The number of entities is not very high: about 10 agent categories, with 5 to 10 verbs for each of them. The simple SIMBAD queries only use a fraction of those.

The management of the current context allows the user to reduce typing: partial commands are built into the prompt, and the user has only to complete them. This context consists of an agent name alone, called the 'current agent', or an agent name and a verb, or, eventually, of a complete command: agent verb and parameters. Executing such a complete command requires only a Carriage Return from the user. The current context can be overwritten at any moment by typing another command, or just another request (verb and parameters) to the current agent.

Let us give here a comparison which should help understand the structure of SIMBAD:

> SIMBAD works like a (good!) post office: in a post office, you find positions or counters, and, in each of them, an agent trained to perform certain functions: selling stamps, receiving parcels to send, paying postal orders, and so on. Some of these agents may respond to several requests. Others send you to a colleague to carry out your request.
>
> SIMBAD can be seen as a collection of different agents, each executing specific functions through different commands.

9–6 How to obtain access to SIMBAD?

Apart from a simple terminal, the only requirement for accessing SIMBAD is obtaining an account number from CDS which will be used for invoicing.

The SIMBAD database is hosted on a DECsystem 5400 running Ultrix (the DEC's version of Unix), at the Strasbourg Observatory. Connections are available through public data transmission networks (namely: TRANSPAC in France, DATEX-P in Germany, TELENET in the U.S., *etc*.), and through SPAN (DECNET) and Internet (TCP/IP) networks. SIMBAD may also be accessed, in France, through TELETEL.

In Europe a gateway to SIMBAD is available through the European Space Information System (ESIS) of the European Space Agency at ESRIN, Frascati, Italy. A gateway for Internet and SPAN is available at the Smithsonian Astrophysical Observatory (SAO) (Cambridge, U.S.A.) and is open, free of charge, to the U.S. users thanks to a specific effort of NASA, as a complement to the Astrophysics Data System (ADS).

It is to be noted that E-mail networks which do not allow remote log-in, such as EARN/Bitnet, are not suitable for an interactive access to SIMBAD.

SIMBAD is a charged service: the users contribute by covering partly the cost of the resources involved in the database service —roughly an order of magnitude lower than for commercial bibliographical databases (which usually also charge for personnel and long-term costs). A reason for this procedure is that, while providing an international service, the CDS is funded by French national institutions only.

9–7 The users of SIMBAD

To the present date (September 1990) there are some 500 observatories, astronomical institutes, associations, or individuals having an account number on SIMBAD. The figures are rapidly increasing.

The following statistics are based on one year of SIMBAD usage (July 1989 to June 1990): 16,101 interactive sessions were performed (giving a total of 77,000 since the first days of SIMBAD, in 1981). During an average session, 14 objects are retrieved, and 85 bibliographical references are displayed, and the user spends 15 minutes in front of his/her terminal. The updating of the database represents an important part of the usage: nearly 25% of the total computing time is devoted to this task, implying about sixty hours of connection per week.

9–8 Updating SIMBAD: a very large cooperative effort

SIMBAD is continuously growing and kept up-to-date, not only by the Strasbourg CDS staff, but also by many cooperating persons in other institutions. This includes a complete survey of the literature in order to give an immediate access to the current bibliography of each object, and the compilation of published observations by experts in the field, for the measurements section.

The bibliography is updated on a daily basis, thanks to the enthusiastic efforts of the librarians of the Institut d'Astrophysique de Paris (IAP) and of the Observatories of Paris, Bordeaux and Strasbourg (see Carpuat et al., 1984).

Catalogues of measurements (original data and mean values when available), produced by specialists, are integrated into SIMBAD by the CDS staff. Only published data are included, and a reference to the source is given. Collaboration with institutes having specialization in specific fields is then a natural consequence. This is particularly the case for:

bibliography: Institut d'Astrophysique de Paris, Observatoire de Bordeaux and Observatoire de Paris;

photometry: Observatoire de Genève and Institut d'Astronomie de Lausanne;

astrometry: Astronomisches Rechen Institut (ARI), Heidelberg

galaxies: Observatoire de Lyon and NASA/IPAC Extragalactic Database (see chapter 10, this book).

9-9 Future trends

Although SIMBAD can already be considered as a large data base with its 740,000 astronomical objects and 250 Mbytes of compressed binary information, a considerable increase in size is soon to be expected with the arrival of new data gathered by satellites or automated ground-based detectors.

These data will not only consist of new measurements to add to the presently known objects, but also of a huge amount of new objects: the delivery of the Guide Star Catalog for the Hubble Space Telescope (see chapter 7, this book) will increase from one to twenty millions the number of objects in an astronomical database like SIMBAD. Also the storage of images will require capacity for large data volumes. CDS is planning the development of new tools for the efficient management of 2- or 3-dimensional data (extended objects, images), a task which is made easier because of the object-oriented programming environment.

One promising development, in the near future, is the integration of SIMBAD in the European Space Information System (ESIS) project (see chapter 14, this book). SIMBAD is one of the nodes of the ESIS Pilot Project, and CDS is actively contributing to the definition of the ESIS Query

Environment [4]. Discussions are also under way for the integration of SIMBAD as a service of the Astrophysics Data System (chapter 15, this book).

It is planned that SIMBAD be connected, in a now rather short timescale, through such systems, with the observing logs and archives of the main ground and space observatories. A user, connected to the system will then have access to the information contained in each database and archive, possibly using SIMBAD as a main index for cross-identifications. Moreover, he shall be able to query all these databases through the same user interface.

The CDS is also currently working on the development of an expert shell which will provide the SIMBAD managers and the SIMBAD users with 'intelligent' astronomy–specific tools for quality control of the database, as well as for source identification, and integration of new data provided by the past and coming space missions, under the scientific responsibility of mission specialists. The recent developments of hypermedia, expert systems and artificial intelligence techniques (see *e.g.* Parsaye *et al.*, 1989) will be particularly useful to accomplish this task.

Acknowledgements

The new SIMBAD III software has been developed with the support of Institut National des Sciences de l'Univers (INSU) and Centre National d'Etudes Spatiales (CNES), and in collaboration with the Hipparcos INCA team in Observatoire de Paris, Meudon. The fruitful collaboration of *Paris Sud Informatique* (the Computer Center of Université Paris XI, in Orsay), host of SIMBAD during the years 1985 to 1990, is also gratefully acknowledged.

References

[1] Carpuat C., Damgé M., Dubois P., Kirchner S., Lagorce A., Laloë S., Ochsenbein F. & Wagner M.-J. (1984), *Bull. Inf. CDS*, **26**, 83.
[2] Egret D. (1986), *Bull. Inf. CDS*, **30**, 25.
[3] Egret D. & Wenger M. (1988), in *Astronomy from Large Databases*, F. Murtagh & A. Heck (Eds.), ESO Proceedings **28**, 323.
[4] Egret D., Ansari S. G., Denizman L., Preite-Martinez A. (1990), "The ESIS Query Environment", CDS Report, Strasbourg.
[5] ESO, Image Processing Group (1987), TERMWINDOWS, A Terminal Independent Screen Formatting and Window Package, ESO, Garching.
[6] ESO/ST-ECF (1987), PROTEUS, A Generalized User Interface Construction Kit, ESO and ST-ECF, Garching.

[7] Goldberg A. & Robson D. (1983), Smalltalk-80 - The language and its implementation, Addison-Wesley Publishing Company.
[8] Heck A. & Egret D. (1987), *The Messenger*, **48**, 22.
[9] Heck A. (1991), Astronomy, Space Sciences and Related Organizations of the World – ASpScROW 1991, *Publ. Spéc Centre Données Strasbourg*, **16**.
[10] Jaschek C. (1988), *Data in Astronomy*, Cambridge University Press, p. 134.
[11] Ochsenbein F., Bischoff M. & Egret D. (1981), *Astron. Astrophys. Suppl.*, **43**, 259.
[12] Ochsenbein F. (1982), in *Automated Data Retrieval in Astronomy*, C. Jaschek and W. D. Heintz (Eds.), I.A.U. Colloquium **64**, Dordrecht, D. Reidel Publishing Company, 171.
[13] Ochsenbein F. (1990), *Catalogue Acronyms and Abbreviations in SIMBAD*, CDS, Strasbourg.
[14] Parsaye K., Chignell M., Khoshafian S., Wong H. (1989), *Intelligent Databases*, Wiley Ed.
[15] Wenger M. (1987), *Bull. Inf. CDS*, **32**, 23.
[16] Wenger M. (1989), *Space Information Systems Newsletter*, **1**, 1.
[17] Wenger M., Bonnarel F., Lesteven S., with the collaboration of Watson, J. M. (1990), SIMBAD User's Guide, CDS, Strasbourg.

10

The NASA/IPAC Extragalactic Database

G. Helou, B. F. Madore, M. Schmitz, M. D. Bicay, X. Wu, and
J. Bennett
Infrared Processing and Analysis Center
Jet Propulsion Laboratory
California Institute of Technology
Pasadena, CA 91125, USA

10-1 Introduction

The NASA/IPAC Extragalactic Database (NED) is a research tool in the form of a central archive which accumulates published data, is organized for fast, flexible retrieval, and is accessible to all astronomers. This service was designed in response to an increasing need driven by ever larger data sets collected in surveys, by the exponential growth in the volume of the technical literature, and by a greater tendency among astronomers to take a multi-spectral approach to the study of astrophysical problems.

NED consists of two main components: a computer-based data system supporting an object-oriented view of extragalactic astronomy, and a convenient user-oriented interface providing access to the data. Access is primarily by researchers or librarians who depend on electronic network connections to reach NED at its physical location at the Infrared Processing and Analysis Center (IPAC) on the Caltech campus in Pasadena, California (USA).

10-2 Context

Astronomical catalogs have always played a crucial role in research, be it observational, empirical or analytic. Catalogs are an invaluable tool to statistical astronomy, and a permanent source of surprises, of new classes of objects hiding in lists of "normal" objects; the successes of moderate catalogs along these lines have usually motivated the creation of bigger and better ones.

Lists of galaxies were being assembled as part of the early surveys for diffuse nebulae even before the term "extragalactic" had been coined. The advent of photographic sky surveys lead to the first catalogs of galaxies with tens of thousands of entries (*e.g.* the CGCG by Zwicky *et al.*). The number of galaxies in catalogs has grown exponentially with time, from about thirty at the time of Messier to the millions being generated today by plate-scanning machines like COSMOS. At the same time, specialized catalogs of galaxies (interacting, low surface brightness, *etc.*) and of quasars have also been multiplying.

Over the past three decades, advances in technology, especially for space-based astronomy, finally made it possible to observe in spectral windows other than the visible range. The exploration through these new windows again turned to sky surveys, resulting in catalogs of radio, x-ray, or infrared sources. These surveys renewed the interest in optical catalogs, for the comparison of old and new yielded important clues to the nature of the new sources.

Fuelled by the same advances and pressures, the astronomical technical journals have been witnessing their own explosive growth in the publication rate. Abt (1988) estimates that the number of articles related to galaxies stood at 1,500 for the publication year 1985, and that it is doubling every eight years. Galaxies and cosmology are the fastest growing among eleven sectors of the astronomical literature surveyed by Abt.

There is little doubt that the trend towards more data will continue, driven by NASA's Great Observatories (Hubble Space Telescope, Gamma-Ray Observatory, Advanced X-ray Astronomy Facility, Space Infra-Red Telescope Facility), other space missions such as ESA's Infrared Space Observatory, and ground-based instruments which are constantly improving in sensitivity, resolution and efficiency. This trend makes it harder for individual researchers to keep up with their specialties, in terms of being aware of new data as well as tracking new ideas. These difficulties reflect the dual challenge presented by the explosive data growth: dealing with the sheer volume, but also inter-connecting intelligently the huge variety of information available. In response to this challenge, attempts have been made to consolidate the published results into reference catalogs (*e.g.* the effort by de Vaucouleurs *et al.*, 1976) which culminates in the publication of RC3 in 1991), or more specialized compilations of data such as the Palumbo *et al.* (1983) Catalogue of Radial Velocities, or the Huchtmeier and Richter (1989) General Catalog of HI Observations.

In the early 1980's the Centre de Données Stellaires of the Observatoire de Strasbourg introduced a new approach to dealing with data, namely a computer database service called SIMBAD (Set of Identifications, Measurements and Bibliography for Astronomical Data), which focused then on stellar data. The innovation was in the exclusively electronic archiving and interrogation, and the continuous updating of the contents. Since then, many similar systems and services have been set up, making up collectively a significant element of the astronomy research environment, as this book amply demonstrates.

In 1987, a fortuitous confluence of interests and opportunities, and the inspiring success of SIMBAD, prompted a number of IPAC astronomers to propose to NASA to establish an archive of published extragalactic data and bibliography, to keep it up to date, and to make it accessible to the astronomical community. IPAC as an environment was well suited to such an enterprise in terms of infrastructure, expertise, and experience with catalogs (*e.g.* Cataloged Galaxies and Quasars Observed in the IRAS Survey 1985). The project having been peer-reviewed and funded, work on the design and implementation started in June 1988. The first release of the NASA/IPAC Extragalactic Database was announced at the 176th Meeting of the American Astronomical Society at Albuquerque in June 1990. NED is now available as a research tool, freely accessible over the electronic networks, without need for prior arrangement, and at no cost to the user. As will become clear in this article, the present capabilities of NED are only part of what it will eventually offer.

10–3 Data contents

NED is an object-oriented database, meaning that it organizes all information around individual astronomical objects as opposed to leaving this information stored in catalogs or compilations. In database management terminology, the organization of NED follows the relational database model. It is thus built around a set of tables called the Object Directory; this is the focus of the database structure, and has additional tables connected to it, usually via object identifications. These tables may contain bibliographic references, notes from various sources, copies of original catalogs, published data of arbitrary description, pointers to observing logs and raw data archives, or other types of information to be added in the future, such as images.

3.1 Object Directory

The Object Directory is the cornerstone of the database, the master list of objects recognized by NED, along with their positions and basic attributes (more about this in section 3.2). It was created by establishing thorough cross-identifications among major astronomical catalogs, and it continues to grow by the addition of lists and catalogs of extragalactic interest. As of this writing (September 1990), the Object Directory contains more than 82,000 objects which account for about 200,000 names, most of which originate in just a dozen catalogs. Table 10-1 shows 23 catalogs with more than 200 members each, all of which have already been folded into the Object Directory. The abbreviations in the second column of Table 10-1 are standard nomenclature in NED, and are used in what follows to refer to these catalogs. The number in column 3 is for all entries in the catalog; this is sometimes larger than the number of objects that appear in NED because some of the catalogs include Galactic objects.

Many other, shorter lists of objects have also been folded in (*e.g.* the Palomar-Green list of quasars (Schmidt and Green 1983), or the Haro (1956) blue galaxies with emission lines). Newly discovered objects are continually added as they are introduced in the literature (*e.g.* the Schombert and Bothun (1988) low surface brightness galaxies, or the Hickson (1989) compact groups). Naturally, NGC and IC names and about a hundred special names (*e.g.* Sombrero Galaxy or The Antennae) have also been added to the Directory. Next on our list of priorities are the IRAS Point Source Catalog (1988), the Abell Catalog of Clusters of Galaxies (1989), and the Arp-Madore Catalog of Southern Peculiar Galaxies and Associations (1987). They will probably have been integrated into the database by the middle of 1991.

Most modern catalogs consist either of mostly Galactic or of mostly extragalactic sources (such as the PKSCAT90 radio catalog), or they will come with a relatively efficient prescription for distinguishing between the two populations (like the ESO/Uppsala or the IRAS Point Source Catalog). To appear in NED, an object has to have been classified as extragalactic in the literature, or to have a reasonable expectation (say 50%) of being extragalactic based on its intrinsic properties. Prime examples of the latter would be a radio source with a synchrotron-like spectrum, or an infrared source with galaxy-like colors, if they are far from the Galactic plane. Given this prescription, NED will unavoidably harbor objects belonging to the Milky Way; we find their inclusion preferable to erring in the other direction, and excluding extragalactic sources. Objects re-classified after having been mistakenly called extragalactic will remain in NED just to signal that switch.

Catalog Name	Abbreviation	Number of Entries
Catalog of Galaxies and of Clusters of Galaxies	CGCG	29,418
Morphological Catalog of Galaxies	MCG	29,003
ESO/Uppsala Catalog of Galaxies	ESO	18,438
Uppsala General Catalog of Galaxies	UGC	12,921
Parkes Catalogue 1990	PKS	8,263
Southern Galaxy Catalog	SGC	5,481
Fourth Cambridge Radio Catalog	4C	4,843
Hewitt & Burbidge Optical Catalog of Quasars	[HB89]	3,700
Fornax Cluster Catalog	FCC	2,678
Virgo Cluster Catalog	VCC	2,096
Markarian Emission-line Objects	MRK	1,515
Slezak *et al.* Coma Supercluster Galaxies	[SMB88]	1,199
Fairall Southern Compact & Bright-Nucleus Galaxies	FAIRALL	1,185
Vorontsov-Velyaminov Interacting Galaxies	VV	852
Crampton *et al.* Quasar Candidate List	[CCS88]	755
University of Michigan Emission-line Objects	UM	685
Arakelian Emission-line Objects	ARK	591
Second Byurakan Survey of Emission-line Objects	SBS	521
Third Cambridge Radio Catalog	3C	471
Selected non-UGC Galaxies	UGCA	444
Arp Atlas of Peculiar Galaxies	ARP	338
DDO Low Surface Brightness Galaxies	DDO	243
Kirshner *et al.* Deep Survey of Galaxies	KOSS	213

Table 10–1: Major Extragalactic Catalogs in the Object Directory

Once an object joins the Object Directory, it is labeled by one of the "object types" in Table 10–2 to reflect its discovery method or its nature. Most entries in Table 10–2 are self-evident; an "absorption line source" is one revealed by absorption against a bright continuum source, typically a quasar; "other" sources may be non-existent sources generated by error, or real objects so unusual or rare that they do not warrant a separate type definition, such as isolated intergalactic clouds detected in emission.

Sub-galactic objects within other galaxies are now infrequently found in NED, but will be included eventually in a systematic fashion. For these, we use object type abbreviations based on, and slightly simplified from, those used by SIMBAD (*e.g.* "SN" for supernova, "PN" for planetary nebula, *etc.*). The same object type abbreviations apply to objects within the Milky Way, except that they are prefixed by an exclamation mark to emphasize their Galactic location.

The NED team has exercised more freely its scientific judgment in the construction of the Object Directory than in the compilation of other data. The merger of a new catalog into the Object Directory is conducted with great care, to ascertain that the various names and detections (at different wavelengths) of the same object are correctly identified and grouped together. This merger operation starts with a search of the Object Directory near the position of every new object to find positional matches. If any are found, the names and basic data carried by the matching objects are then compared to confirm the match, or to resolve ambiguities as needed. Well-researched identifications that come as part of a new catalog are used in the name comparison. When resolving conflicts or ambiguities, data are traced back to the original sources whenever possible, and eventually the Palomar Sky Survey or one of the ESO sky surveys are examined. Such cases often end up being signaled by an "essential note" (see section 3.2 below) within NED.

For each object, the Directory contains positional data, names, "basic data" (described in detail in section 3.2 below), and a "preferred object type", *i.e.* the most useful description of this object by one of the categories in Table 2 (*e.g.* "galaxy" or "quasar" is preferred to "infrared source" or "ultraviolet excess source").

Much care goes into the collection of positions into NED, and they are continually over-written by more accurate values as they become available. Positions are stored internally in the J2000 system, along with their uncertainty, and a reference to their origin. All published positional uncertainties are transformed to a representation as a 95% confidence ellipse, whose semi-major and semi-minor axes and position angle are kept. We also store Galactic coordinates, and total Galactic extinction in the blue at the position, derived from the Burstein-Heiles (1978) reddening maps. The various names by which each object is known are stored in uniform NED formats, *e.g.* 4C +00.30 or UGC 00299. Each name is associated with an object type (from Table 10–2) which reflects the kind of survey in which the name originates. When all of this information is displayed, the user can see that UGC 12699 is a galaxy which was detected as an ultraviolet excess source (called MRK 0538), as an emission line source (UM 167), and as an infrared source (IRAS 23336+0152).

3.2 Basic Data

This is the set of data which are most essential to the description of the object at hand. There is a different set of attributes for each object type, but each object has only one set of basic data, corresponding to its "preferred object type". For example, a galaxy is characterized by an optical

The NASA/IPAC Extragalactic Database

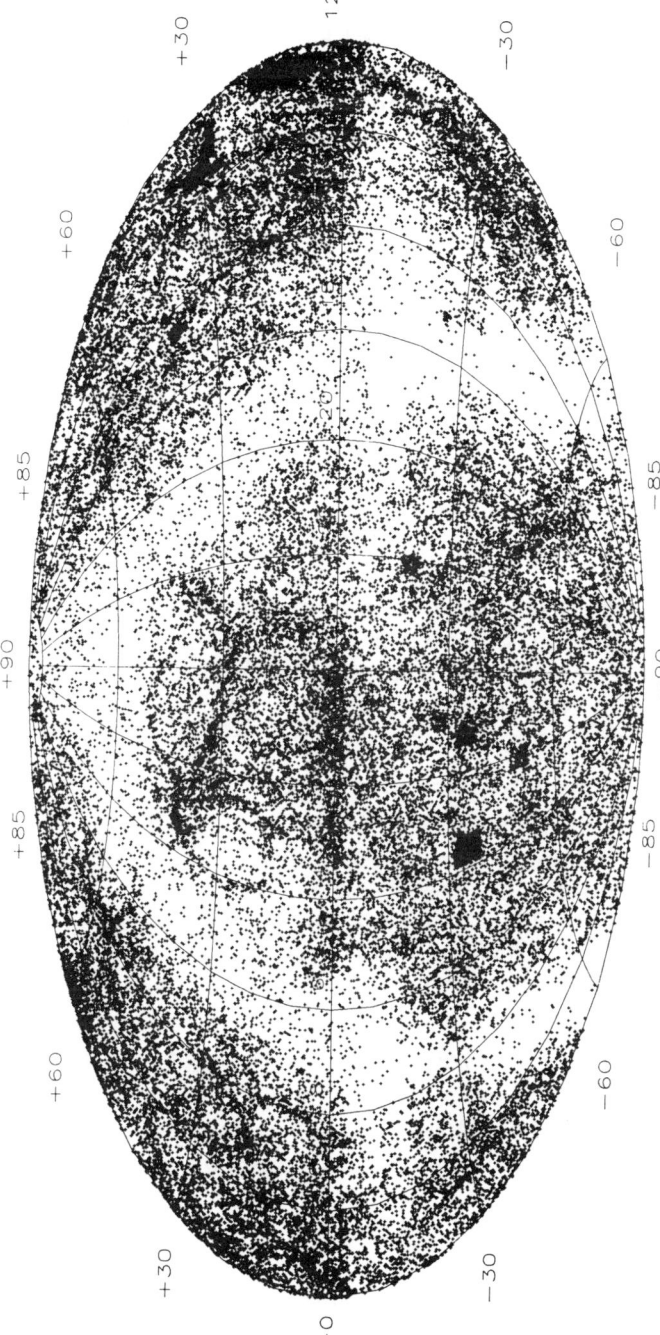

Figure 10–1: Aitoff projection of the sky distribution of 82,791 NED objects plotted in equatorial coordinates, with the origin at the center of the plot, and Right Ascension increasing towards the left. The crowded square area below and left of center is the *Fornax Cluster Catalog* (FCC). The filamentary structure North of center corresponds to the Perseus–Pisces super-cluster. The Coma and Virgo clusters are evident at the extreme right of the plot.

GClstr	Cluster of Galaxies
GGroup	Group of Galaxies
GTrpl	Triplet of Galaxies
GPair	Pair of Galaxies
G	Galaxy
QSO	Quasi-Stellar Object
RadioS	Radio Source
IrS	Infrared Source
EmLS	Emission Line Source
AbLS	Absorption Line Source
UvES	Ultraviolet Excess Source
XrayS	X-ray Source
Other	Unusual object type
PofG	Part of Galaxy

Table 10–2: Extragalactic Object Types in NED

magnitude, a major and minor diameter, a morphological description, and a radial velocity or a redshift; a radio source by a flux density, the radio frequency of that measurement, a spectral index in the vicinity of that frequency, a size, and a morphology (*e.g.* head-tail); a quasar by an optical magnitude, a description (*e.g.* BL Lac or radio quiet), and a redshift.

These basic data are treated as indicative values only, in the sense that they originate in many different sources (which are not explicitly identified in the database), and no attempt has been made to place them on a uniform scale. The main sources are catalogs and compilations, with the more accurate data sets favored, and the larger ones preferred at comparable accuracy. More rigorously defined and referenced data collection into a different segment of the database will take place as described in section 3.6 below.

Finally, "essential notes" generated by NED are attached to some objects (1,100 at present) to point out significant facts, such as an erroneous identification, unique property, or special relation to another object; these notes are always displayed when the object is accessed by a user.

3.3 Bibliographic References

This segment of the database comprises the information providing the most direct connection between the objects and the literature. It consists of pointers indicating the existence of useful information on a certain object in a given publication. In addition, the following data are kept for each publication: a reference code, the publication name, the year, the volume and page numbers, the full title, and the list of authors, with the last two entries limited to 160 characters.

The bibliographic references in NED derive from two main sources. Starting in 1988, one member of the NED team (M. S.) has been reading systematically five major journals to identify papers presenting meaningful new information on extragalactic objects, as well as papers of extragalactic interest in general. The journals covered are *Astronomy and Astrophysics, The Astronomical Journal, The Astrophysical Journal, Monthly Notices of the Royal Astronomical Society,* and *Publications of the Astronomical Society of the Pacific,* along with the associated *Letters* and *Supplements.* This segment of the literature has yielded about 25,000 pointers per year. The second source of bibliographic data is the SIMBAD project, which has kindly provided all of their references to extragalactic objects up to 1989, and will be providing updates on a regular basis. These pointers are based on a more extensive search of the astronomical literature conducted since many years by the librarians at the Institut d'Astrophysique de Paris, and which has produced, for extragalactic sources, systematic coverage starting in 1983, and sporadic coverage going back to 1917.

As of this writing, the database contains about 100,000 pointers linking some 25,000 objects to over 9,000 different publications. Bibliographic pointers to the five journals scanned by NED are typically available on-line about one month after the corresponding issues have appeared in print. New objects introduced by a journal article tend to require more time before integration into the database.

3.4 Notes

Almost every catalog published has a wealth of data appearing as notes on individual objects, usually in an appendix that does not get circulated in computer-legible form, and therefore remains largely untapped. Many journal articles also contribute such notes which are all too easily overlooked. NED has made a special effort to make notes available. In some cases, this has entailed digitizing the printed material for the first time (*e.g.* the Hubble Atlas of Galaxies (Sandage 1961)). In other cases, in particular the ESO/Uppsala and the SGC, the shorthand used in the notes was translated into regular English before the notes were included in the Database; and in the case of the MCG or the Arakelian list (1975) we have had the contents of the notes translated from Russian to English before digitization. As of this writing, the database contains almost 20,000 individual notes, derived mainly from a dozen sources, amounting to 3.5 MBytes of data. The notes are retrievable by query on the object name, and are stored along with a reference to their source, and the particular object name used by the author of the note.

3.5 Literal Catalogs

The major catalogs (such as those listed in Table 10–1 above) are the source material from which NED assembles the Object Directory, and as such need to be accessible in their original form. NED provides that access through the "literal catalogs". The user will be able to view the catalog entries as nearly as possible in their original printed appearance. This capability is still being implemented as of this writing; we now expect to include up to twenty literal catalogs, mostly from Table 10–1, chosen to have the most unique and useful data contents.

3.6 Published Measurement Data

One of the original goals of NED is to collect and store information about all new extragalactic data appearing in the five major journals mentioned in section 3.3 above. NED would carry fully those data that can be expressed in a few numbers, and have clear descriptions of those which cannot, such as spectra and maps. This is clearly an ambitious goal, and is likely to be implemented in a modified or abridged form.

As currently designed, the storage of these data entails the creation of "data frames" containing all the information normally necessary to characterize and make use of each reported measurement. This information includes the essential observational parameters such as wavelength, aperture or beam size, and time of observation, as well as an identification of the telescope and detection equipment used. A dozen varieties of data frames are envisaged, most notably photometric, kinematic, spatial, or derived quantities (such as dynamic mass), and classification or membership data. Photometric quantities for instance would include line fluxes, flux densities, surface brightnesses, and related constructs such as maps, spectra, aperture photometry or radial profiles. A survey of a hundred articles selected at random from the five major journals scanned by NED indicated that the typical article generates about fifty photometric data frames, whereas 3% or so of the papers generate more than a thousand such frames each. Based on that survey, we project 100,000 data frames per year for the bulk of published extragalactic measurements. This estimate points out the enormity of the task, which may not be achievable without assistance from groups presently involved in compiling data, or from the authors of papers in the publication cycle.

3.7 Archival References

A question often asked by researchers is whether a given object was ever observed by a given facility, either as a primary target, or coincidentally as it happened to fall within the field of view. Several facilities (*e.g.* the International Ultraviolet Explorer, and the Very Large Array of the National Radio Astronomy Observatory) already make their observing logs publicly available in computer-legible form (see corresponding chapters in this book). The observing histories of missions such as the Hubble Space Telescope or the Infrared Space Observatory are expected to be particularly valuable and popular resources. NED will integrate into the database such observing logs, and make it possible to search them for arbitrary positions.

This service is expected to reduce the amount of unnecessary duplication in observations, and help locate and acquire data from the archives of space missions.

3.8 Abstracts

When a paper is recognized as relevant to NED in the course of scanning the five major journals (see section 3.3 above), the abstract of that paper is digitized and stored as text, so it can be browsed by users during a NED session. The abstract collection includes not only articles that contribute original data, but all articles of interest to extragalactic astronomy, including reports of theoretical studies, modeling, or empirical analysis. This effort has resulted in 1,051 abstracts for 1988, and 1,140 for 1989, requiring 2.5 MBytes of storage per year of literature.

At present, it is possible to retrieve an abstract only if it is connected to an object by a pointer. We are planning however to implement a more general access method to the abstracts based on searches by topic or key words. Such a capability will represent an added dimension to NED, as it opens a new window on the extragalactic literature which is independent of individual objects and of specific types of data.

10-4 Functions

The users' view of a database is defined by their ability to retrieve data from it. At this time the database provides data only in relation to specific objects, so that locating objects is the primary mode of access (discussed in section 4.1 below), followed by the retrieval of various types of data pertaining to these objects (section 4.2). Ancillary functions provided by NED include a coordinate transformation utility, and the transmission to the user by electronic mail of an ASCII file containing all data retrieved during a session, excluding abstracts.

In addition to these functions, there is an elaborate collection of software tools, largely invisible to the user, whose purpose is to populate NED with new objects and data and update existing information, while preserving data integrity by checking for consistency and maintaining traceability of the modifications.

4.1 Searching for objects

Users can locate objects by specifying a name, or a vicinity to search in. Substantial latitude is allowed in entering names, since the input is interpreted and cast into the standardized formats used by NED for internal storage (more about this in section 5 below). A search by name (*e.g.* VV 136) will normally return (unless otherwise selected by the user) all objects that carry the specified name plus a suffix (*i.e.* VV 136, VV 136a, VV 136b, and VV 136c). Search by vicinity returns all objects lying within a circle, whose radius and center are specified by the user. Such searches can be centered on an object specified by name, in which case the occurrence of suffixed versions of this name are ignored. If the center is specified by position, it can be given in any of four coordinate systems: Equatorial (and here any year can be selected for the equinox), Ecliptic, Galactic, or Super-Galactic.

We are planning for a capability to search by constraints, which will allow users to find objects satisfying a set of conditions, such as "all quasars at Galactic latitudes between 45°and 50°".

Each one of those searches returns a list of objects, along with the corresponding contents of the Object Directory and basic data (section 3.1 and section 3.2 above), and with the number of bibliographic references (section 3.3) and the number of notes (section 3.4) available for each object, thus presenting the user with a summary status of the object within NED. The interface then allows the user to display this list, or examine in detail individual objects within the list. Further information about each object is retrieved by specific requests, as detailed in the next section.

4.2 Retrieving data

Data retrieval requests presently in operation correspond to those data structures already in place described in section 3 above. One can retrieve notes and bibliographic references for specific objects; once a reference has been found, its abstract may be displayed. Similar functions for literal catalogs, archival references and published measurements will be activated as the data are added to NED.

Requests for notes and references can be issued to follow up on an object identified in a previous search for objects, or can be formulated *ab initio* by specifying an object name. The second mode offers (as an option) the advantage of an "extended" search, which starts by locating other objects which have a name root in common with the name specified, then retrieves all references to all objects thus related to the object of interest. Take for instance the case of the Hickson Compact Group HCG 23; it consists of five galaxies called HCG 23A through HCG 23E. NED contains notes pertaining to, and references linked to, only some of the five individual members. A search for articles strictly relevant to HCG 23 does not reveal papers linked to individual member galaxies, whereas an extended search will return all references to HCG 23 as well as HCG 23A through HCG 23E. More useful yet, the same all-inclusive list will be returned by an extended search on NGC 1216 which happens to be HCG 23C.

10–5 Interface

The user's view of a database is strongly colored by the ease with which one interacts with it. This ease is defined by several factors, including: the amount of learning needed before sessions become productive; the degree of overlap between the specific questions a user has and those the database can answer; the power and flexibility available to the user in formulating a query; and the convenience in the mechanics of submitting a search and collecting the results. While some of these factors are determined by the internal structure of the database, they can be modified by, and in the end depend mostly on, the user interface to the database.

In line with those perceptions, the NED interface is conceived as an interpreter that enables an astronomer or a librarian to use the database without learning its jargon. One of its main goals is to make a user's first session a productive one in terms of obtaining the sought-after data, when the user knows nothing of the details of data organization or interface mechanics.

NED is designed for remote use via electronic networks, and may be accessed by setting up a network connection to IPAC on either Internet or SPAN. There is no charge for logging into the service or retrieving data from it. From Internet, one uses the command: "telnet ipac.caltech.edu"; from SPAN, one uses "set host IPAC". The absolute address of IPAC on SPAN is 5.857. Once connected to IPAC and prompted for a "login", the user should respond with "NED"; no password is needed. The interface requires a VT100 terminal, or a software emulation of such a terminal, to function properly.

In the present implementation, users control the NED session by selecting options and entering data within the context of a menu tree; the interface presents itself as a screenful of text for each node in this tree, with a standard screen format maintained throughout the session. This format consists of three boxes: the one at the top displays available options, while the one at the bottom displays commands, and the middle box serves for input and output. Options are displayed as abbreviated descriptions of the functions they serve. As an option is selected (typically with a single key stroke), a new screen is displayed (presenting new options), a data input screen is activated, or a data output sequence is initiated. The commands (all of which are always displayed with explanations in the lower box) perform simple, general purpose tasks such as terminating input or output, moving to a higher-level menu, or ending the session.

A NED session thus involves selecting options until the desired input panel is reached; once the various fields in the selected panel have been filled out, a search defined by the input data is submitted, and the interface is deactivated until the search concludes. The resulting data are then available to the user for display in various output screens, until a new search is requested, causing the new results to overwrite the previous ones. All resulting data returned during a session (except for abstracts) are written to a file which is sent by electronic mail to the user, if so requested at the end of the session.

An essential feature of the NED interface is self–documentation. In the first place, a NED session is steered at every step by a choice among options expressed in astronomical jargon, instead of being driven by an interrogation language the user must learn. Secondly, help functions are available at several levels throughout a session: (1) A brief explanation for every menu item or option, and for every input field on any screen can be obtained by pointing the cursor at that item or field and entering "control-H". (2) A longer explanation of the functions and usages within every screen can be displayed by selecting the "HELP" option within that screen's menu. (3) An introduction and overview, a general tutorial, recent news, and a glossary are also available as menu options in the first screen ("Main Menu") offered by NED.

Similarly extensive documentation is available on output; data fields are labeled in whatever detail is necessary, and abbreviations are avoided as much as possible. On the screen displaying names of objects (cross-identifications) for instance, the user can ask for information on a given name, and get a brief description and a bibliographic reference to the catalog in which the name originates.

As the NED interface is accepting data from a user, it checks them to the extent possible for ambiguities, conflicts or errors; more significantly, it attempts to interpret them, and echoes its interpretation to the user for verification. These verification capabilities translate into less restrictive formats for the user to adhere to; for instance, the right ascension field will accept any of "6h05s", "06::5", or "6 0 5" as valid input, and interpret it as "06h00m05s". The most versatile and sophisticated interpreter in NED is associated with names of objects, since naming conventions show great diversity across the literature. Given some user input, it generates responses of which the following are typical examples: the input is uniquely recognized, and rewritten in the standard NED internal format in preparation for querying the database; the input is ambiguous, so a number of possible interpretations are offered for the user to choose among; or the input is distantly related to one or more naming conventions, so the proper formats for the latter are offered to the user as information. If the input cannot be validated, the response includes a reason for that failure (*e.g.* a numerical value is out of range). This interpreter uses a software module written for NED, and based on the "lex" regular expression processor; it gets the nomenclature guidelines out of a file, thus simplifying the task of updating name conventions recognized by NED.

To underline the importance of the users' suggestions and opinions, every screen offers the user the option of leaving comments on the system. In addition, users are invited to do the same at the end of every session. Comments have included problem reports, possible errors in the data, desirable enhancements to the database, and quality ratings of the service. The comments are reviewed and acted upon almost daily by the NED team.

10–6   Architecture

The NED software resides on two platforms, a SUN3 which supports all the user interface functions and the abstract database, and a CDC CYBER 180 which carries the search functions, and where most of the data are kept and managed. The two computers communicate via a CDCNET local area network, in a way which is completely invisible to the user; this communication is based on a modified *telnet* protocol.

On the SUN3, the main software module which manages the session was written expressly for NED, but relies for screen input and output on a package called YAMM (for "yet another menu manager"), developed by Alan Mazer at JPL. YAMM is responsible for much of the "look and feel" of NED since it handles screen formatting, menu presentation, help functions, and user input by way of simple commands, menu selections, and data entry panels. On the CYBER side, all database management system functions are handled by IM/DM, a commercial software package that follows largely the relational database concept. In spite of the apparent complexity, this architecture offers many advantages in terms of performance, security, and manageability, and makes optimal use of the available platforms.

10–7 Future plans

NED is clearly in its initial growth phase, with several of the originally planned features still under development. Over the next couple of years, most if not all the capabilities mentioned above will have entered into service, including literal catalogs, archival references, and at least a partial implementation of measurement data. Searches for objects by constraints will be possible, as will be access to abstracts by topics of interest. In addition, the existing parts of NED will continue to grow with more catalogs being folded into the Object Directory, or appearing as literal catalogs, more pointers and basic data being entered, and more abstracts becoming available.

In the long term, NED will certainly undergo much evolution, for it was designed with flexibility and ease of upgrades in mind. Enhancements presently envisaged call for improving the user interface, probably by migrating to a more powerful screen manager; for user registration and tracking to respond more readily to individual user preferences; for the ability to deal with batch processing of groups of requests transmitted as files rather than interactively, one at a time. Moreover, the trend towards more powerful computers and communication networks will move NED towards an interface capable of graphic display, and towards the storage, retrieval and browsing of images, spectra, and similar data.

This is a time of transition, with the information revolution finally changing the way astronomers conduct their business. NED is a symptom of that transition, but could also become a catalyst for more change, and lead to new habits in research. NED relies on the existing peer-review system for ensuring data quality, since it draws data primarily from the refereed literature. However, it allows a new mode for tapping that literature (the bibliographic pointers), and will experiment with other modes made possible by the growing collection of abstracts at its disposal.

Acknowledgements

Many people have contributed to NED, and continue to do so; we would like to thank Nian Chiu, Rick Ebert, Alan Mazer, and Maneesh Sahani for their help in the software arena; April Michel, Walter Rice and Carol Lonsdale for help with populating the database; and especially Rosanne Hernandez and Eve Scholey for their dedication to digitizing the abstracts, and John Good for his advice and suggestions. We especially acknowledge the friendly collaboration of the SIMBAD project, Centre de Données astronomiques de Strasbourg. We would also like to thank all the authors who send us computer-legible versions of their data or catalogs. NED is funded by the Science Operations Branch of the Astrophysics Division, Office of Space Science and Applications, NASA. This work was carried out by the Jet Propulsion Laboratory, California Institute of Technology, under a contract with the National Aeronautics and Space Administration.

References

[1] Abell G .O., Corwin H. G. Jr., Olowin R. P. (1989), *ApJS*, **70**, 1 (ABELL).
[2] Abt, H.A. (1988), *PASP*, **100**, 1567.
[3] Arakelian, M.A. (1975), *Soobshch. Byurakan Obs. Akad. Nauk. Arm. SSR*, **47**, 1 (ARK).
[4] Arp, H. (1966), *ApJS*, **14**, 1 (ARP).
[5] Arp H. C. & Madore B. F. (1987), "A Catalogue of Southern Peculiar Galaxies and Associations", Cambridge, Cambridge University Press.
[6] Binggeli, B., Sandage, A., Tammann, G. A. (1985), *AJ*, **90**, 1681 (VCC).
[7] Burstein, D. & Heiles, C. (1978), *ApJ*, **225**, 40.
[8] Caswell, J.L., and Crowther, J.H. (1969), *MNRAS*, **145**, 181 (4C) .
[9] "Cataloged Galaxies and Quasars Observed in the IRAS Survey" (1985), Pasadena, California, Jet Propulsion Laboratory (JPL D-1932) .
[10] Corwin, H. G. Jr., de Vaucouleurs, A., de Vaucouleurs, G. (1985), "Southern Galaxy Catalogue", Austin, Texas, University of Texas (SGC).
[11] Crampton, D., Cowley, A.P., Schmidtke, P.C., Janson, T., Durrell, P. (1988), *AJ*, **96**, 816 ([CCS88]) .
[12] Fairall, A. (1977-88), 7 papers in *MNRAS* (FAIRALL).
[13] Ferguson, H. C. (1989), *AJ*, **98**, 367 (FCC).
[14] Gower, J.F.R., Scott, P.F., Wills, D. (1967), *MmRAS*, **71**, 49 (4C).
[15] Haro, G. (1956), *Bol. Obs. Tonantzintla y Tacubaya* 2, **14**, 8 (HARO).
[16] Hewitt, A., Burbidge, G., (1989), *ApJS*, **69**, 1 ([HB89]).
[17] Hickson, P, Kindl, E., & Auman, J.R. (1989), *ApJS*, **70**, 687 (HCG).
[18] Huchtmeier, W.K., & Richter, O.-G. (1989), "A General Catalogue of HI Observations of Galaxies", New York, Springer-Verlag.
[19] IRAS Point Source Catalog, Version 2 (1988), Joint IRAS Science Working Group, Washington, DC, US GPO.

[20] Kirshner R. P, Oemler A., Jr., Shechter P. L., Shectman, S. A. (1983), *AJ*, **88**, 1285 (KOSS).
[21] Lauberts, A. (1982), "The ESO/Uppsala Survey of the ESO(B) Atlas", Garching bei München, European Southern Observatory (ESO).
[22] Markaryan, B.E. et al. (1967-1981), 15 papers in *Astrofizika* (MRK).
[23] Markaryan, B.E. et al. (1983), *Astrofizika*, **19**, 639 (SBS).
[24] McAlpine, G.M. et al. (1977-81), 5 papers in *ApJS* (UM).
[25] Nilson, P. (1973), "Uppsala General Catalogue of Galaxies", *Acta Universitatis Upsaliensis, Ser.* V:A vol. 1, Uppsala, Reg. Soc. Scient. Ups. (UGC).
[26] Nilson, P. (1974), "Catalogue of Selected non-UGC Galaxies", Uppsala Astronomical Observatory Report 5 (UGCA).
[27] Palumbo, G.G.C., Tanzella-Nitti, G., and Vettolani, G. (1983), "Catalogue of Radial Velocities of Galaxies", New York, Gordon & Breach.
[28] Parkes Catalogue (1990), the Southern Radio Source Database, ver. 1.01, compiled by A.E. Wright and R.E. Otrupcek. Magnetic tape and preprint, Australia Telescope National Facility (PKS).
[29] Pilkington, J.D.H., & Scott, P.F. (1965), *MmRAS*, **69**, 183 (4C).
[30] Sandage, A. (1961), "The Hubble Atlas of Galaxies", Washington, DC, Carnegie Institution of Washington.
[31] Schmidt, M., and Green, R.F. (1983), *ApJ*, **269**, 352 (PG).
[32] Schombert, J.M., and Bothun, G.D. (1988), *AJ*, **95**, 1389 ([SB88]).
[33] Slezak, et al. (1988), *A&A Suppl. Ser.*, **74**, 83 ([SMB88]).
[34] Smith, H. E., Spinrad, H., Smith, E. O. (1976), *PASP*, **88**, 621 (3C).
[35] van den Bergh, S. (1966), *AJ*, **71**, 922 (DDO).
[36] de Vaucouleurs, G., de Vaucouleurs, A., & Corwin, H.G., Jr. (1976), "Second Reference Catalogue of Bright Galaxies", Austin, Univ. of Texas Press (RC2).
[37] Vorontsov-Velyaminov, B. et al. (1962-74), "Morphological Catalogue of Galaxies", in 5 parts, Moscow, Moscow State University (MCG).
[38] Vorontsov-Velyaminov, B.A. (1959), Atlas and Catalog of Interacting Galaxies, Sternberg Institute, Moscow State University (VV).
[39] Vorontsov-Velyaminov, B.A. (1977), *A&A Suppl. Ser.*, **28**, 1 (VV).
[40] Zwicky, F. et al. (1961-68), "Catalogue of Galaxies and of Clusters of Galaxies", in 6 volumes, Pasadena, CA, California Institute of Technology (CGCG).

11

The ESO Archive Project

François Ochsenbein
Image Processing Group
European Southern Observatory (ESO)
Karl–Schwarzschild Straße 2
D – 8046, Garching bei München, FRG

11-1 Introduction

The high quality of the observations made with the telescopes of the new generation (see *e.g. The Messenger* 59, page 14) is a very strong motivation for a systematic archiving of the data acquired at the telescopes of the European Southern Observatory (ESO), located in the *Norte Chico* at La Silla, Chile. The other —more classical— reasons to archive data obtained at ESO telescopes are: (a) to keep an historical record of observed objects, for later analysis of possible long–term variabilities   (b) to reuse data for other scientific purposes and therefore avoid the duplication of observations and   (c) to allow research based on the accumulated material, *e.g.* statistical studies, evolution of an object over several years, including studies on the evolution on the instruments.

11-2 Raw versus Calibrated data

A distinction is generally made (Jaschek, 1988) between *raw data* —basically what comes out of the detector— and *calibrated data* where the *known* instrumental biases are removed and the data are expressed in physical units. A further step, normally achieved by the observing scientist, leads to *reduced data*.

Raw data —*i.e.* pixel values associated to the various telescope and instrumentation setting parameters called the "header" or *data descriptor* —

will likely be difficult to interpret by the next generation of scientists, when the technical expertise required for operating and managing the instrument fades away; on the other hand, the calibration procedure is regularly refined during the instrumentation lifetime, leading to more accurate data derived from the same raw data input. The example of the several iterations performed on the IUE low dispersion spectra (Cassatella *et al.*, 1988; see also chapter on IUE, page 35, this book) strongly suggests to keep the actual raw data to allow for later reprocessing.

It could be envisaged to store raw data *and* the calibration procedures; but how exactly to archive a 'procedure' is not well defined, due to the rapid evolution of software and software tools (what would MIDAS commands look like ten years from now ?).

The conclusion could be that raw data must be archived for at least the lifetime of the instrument. The final archive —the one which stays for the future generation— would normally only consist in (hopefully well) calibrated data.

11–3 Archiving at ESO

The Archiving Policy defined at ESO (Van der Laan, 1988) grants a proprietary period for scientific data of one year after termination of the observations; this proprietary period may be extended on request in special cases. Non–scientific data, *e.g.* calibration data and catalogues of observed targets, will normally become public immediately after the end of the observations.

Systematic archiving will first start with the New Technology Telescope (NTT), which will be available for scientific observations in the beginning of 1991. The amount of data coming out of the NTT will be rather large (presently 1024×1024 CCD detectors are used); it is estimated to be between 0.2 and 0.5 Gbytes/day, *i.e.* of the order of 100 Gbytes/year. This amount will grow rapidly: the usage of 2048×2048 CCD frames is scheduled for the near future, and the data coming out of the other telescopes will also be archived. The amount of data that will be produced in the future by the Very Large Telescope (VLT) is difficult to predict right now; with the four telescopes, and the large CCDs, it should be 64 times larger than the NTT—therefore in the (10^{13}) bytes/year range.

Since raw data are archived, as discussed in the previous section, *calibration data* are an important part of the Archive. It is also planned to make regular observations of standard objects, for data reduction purposes as well as for a better knowledge of the evolution of the instrument.

11–4 Data Format

Each archived raw data set includes the image stored as a two-dimensional image, together with a *data descriptor* specifying when, where, and how the data were acquired.

The FITS format (Wells *et al.*, 1981; Grosbøl *et al.*, 1988; see also related chapter, page 253, this book) allows to store binary data associated to a complete description in ASCII form of the data. This format was chosen for the Archive —even if it is not quite efficient in terms of storage requirements— because it has the main advantages of being easily readable (the data descriptor is in plain ASCII), computer–independent (the Archive can therefore be moved to a different computer without reformatting the data), and widely known and used in the astronomical community: all major data reduction packages are able to read and write FITS images. This format was also recommended by the International Astronomical Union (IAU) at the Patras General Assembly [5], including extensions [6].

In the data descriptor, which is stored as the FITS header, an observation is described by means of keywords followed by a numeric or character value. Basic FITS keywords defined in the original paper (Wells *et al.*, 1981) are used for very general descriptions: BITPIX, NAXIS, NAXIS1, NAXIS2 provide the size of each image; TELESCOP, INSTRUME, OBSERVER designate the instrumentation and the observer's name; DATE-OBS, EXPTIME specify when the observation was performed, and the exposure duration; a MJD-OBS keyword (modified Julian date, *i.e.* $JD - 2400000.5$) is added to have a more accurate date and time stamp; RA, DEC provide the apparent position of the telescope at the date of observation; OBJECT is the original designation of the target, and TARGET is a "standard" object identifier homogenized over the whole Archive. The observer's comments are recorded under the COMMENT keyword, and finally the part of the telescope log concerning the reported observation is copied to records starting with the LOG-ESO keyword.

The original FITS scheme does not however easily allow a complete description of all observing peculiarities and instrument setup for a complicated instrument like EMMI, the ESO MultiMode Instrument available on the NTT, unless cryptic keywords like ESA2W3F (for a filter on Arm 2, Wheel 3) are used; this would clearly outstand one of the arguments of using FITS for its understandability. An accurate description of the technical parameters is achieved via a *hierarchical scheme* (see chapter on FITS, this book) starting with the unique keyword "HIERARCH", followed by ESO, and then a keyword that specifies one of the five categories detailed below; further keywords designate the subsystems, if any, and finally *the parameter*. The categories are:

1. GENeral is related to the observation run (project identification), and to the classification of the observation, as a *scientific* or a *calibration* observation or the observation of a *standard* object.
2. TELescop describes the telescope setting, *e.g.* the focus used.
3. ADApter describes the optical elements (position, rotation angle) of the adapter used.
4. INStrument describes the instrument setting, *e.g.* the filters used, the dispersive system with its orientation, *etc.*
5. DETector describes the parameters of the detector, *e.g.* its name, mode, status, the gain, the temperatures, voltages, *etc.*

The usage of this hierarchical scheme will still allow the presently existing astronomical image processing systems to retrieve the parameters required for basic manipulation of the data; most of the engineering data reported in the deep-down levels of the structure would not have a large impact on the results, but *may* be useful to detect and eventually correct unforeseen effects. The MIDAS software, used at ESO and distributed in more than 100 observatories, will be able to make use of these keywords for data reduction if required.

11–5 The Archive Medium

The large amount of data stored in the Archive requires obviously large capacity storage media. The recent technology produces optical disks and tape cartridges; see *e.g.* Pirenne & Ochsenbein (1990) for a short review of such new media.

The evolution of the capacity of the new tape cartridges, for instance, is really impressive: their density goes up to 30 Mb/cm^3, which is *three* orders of magnitude more compact than the classical 1600-bpi tapes; the comparison of the average cost per Megabyte stored leads to even higher figures. How long the data can be stored on such media without degradation is not yet well tested. Optical tapes, with capacities in the Terabyte (10^{12}) range, are now being tested and will soon be available.

It is generally stated that the WORM (Write Once, Read Many times) optical disks has a lifetime larger than a decade; this is quite compatible with the period during which we want to keep the *raw data*. The optical disks are also attractive for their direct access capability and for the (relatively) low cost of the Gigabyte stored, but no standard for the organisation of the file system on such disks has yet emerged.

A flat file format for WORM disks was however agreed upon between STScI, ESO and ST-ECF (Richmond *et al.*, 1987); its main purpose is to be machine–independent, allowing therefore a transparent access to files on different operating systems. Such a file system associated to an internal FITS format ensures the portability (in time) of the Archive: it is possible to move all data from one computer to a completely different platform without data migration—that is within a few days. A set of routines allowing access to files stored with this format, written in C language and layered on top of the MIDAS operating system routines, will soon be available on request.

Which storage medium will be used still remains an open choice for the ESO Archive. It will likely be a combination of several media, for instance optical (or magnetic) disks for frequently used data (*e.g.* fundamental calibrations), and cartridges for more temporary data. It will also obviously evolve with the data storage techniques.

11–6 The Archiving Process

The data coming out of the telescopes at La Silla will be sent to Garching at regular intervals for archiving. The archiving process itself digests the electronic logfile, checks the consistency of the data, writes the file to the archival media, and updates the catalogue. The checking consists in classifying the observations (as scientific, calibration or standard object) and in verifying the correctness of key parameters such as date and time, observer's name, and of the object identifier.

A test over one year of EFOSC (ESO Faint Object Spectrograph and Camera) observations on the $3.6m$ telescope was performed to ensure the feasibility of the project. The reliability of some key parameters (telescope position, timing parameters, target names) was checked, including systematic comparisons of the telescope position versus the target name using remote program-driven connections to the SIMBAD database (Ochsenbein, 1988). As a result, detailed specifications on the contents and formatting of the output from NTT observations was defined in order to ensure maximal reliability of the archived material: accurate definitions of the parameters to be recorded, their formats, the contents of the electronic log, *etc.*

The recent installation of a digital link with a bandwidth of 64 Kbit/s between the ESO headquarters at Garching and the telescopes will allow a better interaction between the archive and the observer(s): it will enable a possibility of archiving the observation logs almost in real time, and allow the observers on the mountains to access the archive and other on–line data (typically astronomical catalogues, see below) during the observation.
Many other communication problems can also be solved smoothly, such as real-time reports of modifications in the schedule that occur inevitably in the very last moments, avoiding the discrepancies found in the EFOSC

test between the scheduled telescope usage and the observations actually performed.

11-7 The Catalogue of Observations

The Catalogue of ESO observations is mainly built from the *data descriptors* of each observation. It is organised as a set of tables installed in a commercial database, including:

- parameters useful for queries by in-house or remote astronomers: main instrument parameters, telescope position, exposure time, image size, target designation;
- the list of proposals accepted by the Observation Program Committee, with applicant names and full title;
- technical data (instrument descriptions);
- contents of the archived media (archiving processing status);
- status of requests for dearchiving.

The Catalogue may be queried via STARCAT (see the next section); the data displayed on the user's screen are actually a database view merging several of the tables mentioned above.

The catalogue will be accessible for queries at the ESO computer facilities in Garching and in La Silla, and over computer networks (presently SPAN and public networks (X.25), Internet in the near future). Access to non-proprietary data will be possible at ESO Garching, and maybe at ESO-La Silla (depending on the speed of the link); shipment of large data amounts from the ESO Archive are subject to scientific evaluation.

11-8 STARCAT: The user interface

STARCAT is a piece of software developed in collaboration with the ST-ECF and the STScI; it is mainly a user interface that interacts with the database, providing some astronomical facilities to build queries. Details about its structure can be found elsewhere (see *e.g.* Pirenne *et al.*, 1988); it was designed to be *user-friendly* (no need to read a huge User's Manual), and *portable*: operating system (OS) dependencies are insulated in a small set of *OS-routines*, and the connections to the DBMS are interfaced by a set of *STDB-routines*. This allows to move Starcat from one operating system to another, and from a DBMS to another DBMS; both changes were actually performed without any major problem.

STARCAT is *form–oriented*, *i.e.* after having chosen his favourite catalogue (or view), the user sees a form on the screen, and a menu. To generate a query, the user just hits the qualification key, and then navigates in the form; he types in the various constraints in some of the fields where he wants some restrictions, *e.g.* <10 in the *vmag* field to specify bright objects. Once the constraints have been entered, the action of the FindNext key starts the retrieval and the first found record that matches the specifications is displayed in the form. If no constraint was specified, each record of the catalogue is successively retrieved. A special *Coordinate Selection* window pops up if the Center key is hit; constraints on the position (celestial coordinates) in the equatorial, galactic or ecliptic frames can then be entered. STARCAT also allows to compute "on the fly" various parameters that are not in the original tables; the most typical example is the computation of coordinates in several frames and/or epoch/equinox.

STARCAT is now in use at ESO, at the Space Telescope Science Institute (STScI) in Baltimore, at the Canadian Astronomical Data Center (CADC) at Victoria, and the National Space Science Data Center (NSSDC) for quite some years, mainly for access to astronomical catalogues. The feed-back from astronomers using STARCAT at ESO, both in-house and remote users, is generally positive: the system is *simple* to use, which is appreciated by the majority of our users, whereas a few —very few in fact— find Starcat a bit too primitive.

About 40 astronomical catalogues are presently available via Starcat; the list is regularly updated, old versions are replaced with new ones, new catalogues are added, generally received from the Centre de Données de Strasbourg (CDS), but also from other sources. The choice of the catalogues stored in our database mainly reflects the scientific interests of the ESO scientific group and of our visitors. Special care is taken about the documentation of the catalogues (available through the *Documentation* key in the Starcat menu), which as frequently as possible includes the explanation from the original author(s).

Among other catalogues, STARCAT allows access to the ESO–LV Surface Photometry Catalogue (Lauberts & Valentijn, 1989); the calibrated images of the galaxies (28,218 images stored on optical disk) can be retrieved via the *File Handler* (McGlynn & Hunt, 1988), or via MIDAS special commands.

STARCAT also allows a consultation of the lists of preprints and periodicals received at the ESO Libraries. A facility for a transparent access to remote databases like SIMBAD or the IUE Vilspa database was added to STARCAT; the SIMBAD interface was recently modified to accommodate the move to the new version of SIMBAD (see page 79, this book).

Acknowledgements

Developments of the ESO Archive evolved thanks to fruitful discussions and collaborations within ESO. The software was mainly developed with the participation of the Image Processing Group, and the Archive Group of the ST-ECF.

References

[1] Cassatella A., Lloyd C., Gonzales Riestra R. (1988), *IUE Newsletter* **31**, 13.
[2] McGlynn T. A. & Hunt L. K. (1988), ST-ECF O-02 Document, volume IX.
[3] Wells D. C., Greisen E. W., Harten R. H (1981), *Astron. Astrophys., Suppl. Ser.* **44**, 363.
[4] Grosbøl P., Harten R. H., Greisen E. W., Wells D. C. (1988), *Astron. Astrophys., Suppl. Ser.* **73**, 365.
[5] Resolution C1 (1983), *IAU Information Bulletin* **49**, 14.
[6] Resolution B1 (1989), *IAU Information Bulletin* **61**, 10.
[7] Jaschek C. (1988), *Data in Astronomy*, Cambridge University Press, chap. 4.
[8] Lauberts A. & Valentijn E. A. (1989), *The Surface Photometry of the ESO-Uppsala Galaxies*, European Southern Observatory.
[9] Ochsenbein F. (1988), in *Astronomy from Large Databases*, F. Murtagh & A. Heck (Eds)., ESO Proceedings **28**, 429.
[10] Pirenne B. & Ochsenbein F. (1990), *ST-ECF Newsletter* **12**, 11.
[11] Pirenne B., Hunt L., Richmond A., Russo G. (1988), *Astronet Workshop on Databases*, Bologna.
[12] Richmond A., Russo G., Ochsenbein F., Mc Glynn T. (1987), *Proc. Optical 87 Conference*, London, 331.
[13] Van der Laan H. (1988), *The Messenger* **52**, 3.

12

Archives of the Isaac Newton Group, La Palma and Westerbork observatories

Ernst Raimond
Netherlands Foundation for Research in Astronomy
Radio Observatory
P.O. Box 2
7990 AA Dwingeloo, the Netherlands

12-1 Introduction

Historically, maintenance of an archive of observations was not the highest priority of most groundbased observatories. In that respect space observatories like IUE (chapter 5) have set a better example. A mere collection of magnetic tapes containing most of the observations is not considered to constitute an archive in this context. Addition of a suitable index or catalogue enabling an outside user to find observations by their parameters would turn the collection of tapes into a proper archive. This chapter deals with the archives of two observatories where archiving was built into the system from the start.

At present, the British/Dutch Isaac Newton Group of Telescopes on La Palma, Canary Islands and the Westerbork Synthesis Radio Telescope (WSRT) in the Netherlands use the same or similar software for archiving and for accessing the archived data. This justifies describing the two archives together. Within one or two years the British/Canadian/Dutch James Clerk Maxwell submillimeter Telescope and the United Kingdom Infrared Telescope (UKIRT) on Hawaii will also make use of the same archive software.

The Isaac Newton Group of telescopes on La Palma: La Palma, one of the Canary Islands, is the site of an international astronomical observatory, the Observatorio del Roque de los Muchachos del Instituto Astrofisica de Canarias (IAC). The IAC has its headquarters in La Laguna, Tenerife. The Isaac Newton Group of Telescopes (ING) is operated by the Royal Greenwich Observatory on behalf of the British and Dutch astronomical communities. The ING consists of three telescopes: the 1-m Jacobus Kapteyn Telescope, the 2.5-m Isaac Newton Telescope, and the 4.2-m William Herschel Telescope. The Kapteyn and Newton telescopes have been in operation since 1984, the Herschel telescope since 1987.

The Westerbork Synthesis Radio Telescope: The Westerbork telescope (WSRT) is an aperture synthesis instrument, owned and operated by the Netherlands Foundation for Research in Astronomy (NFRA). The telescope is used mainly by the Dutch astronomical community but is open to observing proposals from astronomers in other countries. It has been in operation since 1970.

12-2 Contents and organization of the archives

2.1 The ING, La Palma archive

The data of the ING telescopes are recorded in a structured FITS format on magnetic tape. The structuring consists of grouping the parameters describing the observation, the actual parameters of the telescope, the parameters and settings of the observing instrument and those of the detector into entities within the FITS file. In general the observed data are part of the detector-packet. The observer leaves La Palma with a copy of the original data tape. The original tape, of which the readability has been certified in the copying process, is kept on the island for several weeks in order to minimize the risk of losing data in transfer. Thereafter these tapes are sent to the Royal Greenwich Observatory, Cambridge, England, to be loaded into the archive.

Archiving the La Palma data is performed by a program, which is part of a comprehensive set designed and developed by the Netherlands Foundation for Research in Astronomy in Dwingeloo. It involves checking the format of the data, storing it in a more compact internal format, building an index of all individual entities (packets), keeping an administration of the tapes on which the data was transported from La Palma, maintaining an administration of the archive tapes and constructing and maintaining a Catalogue of Observations.

The archive of ING observations is maintained by the La Palma Users Unit of the Royal Greenwich Observatory and resides in Cambridge, England. It consists of the following components:

1. Observed and ancillary data, stored in a well defined format on magnetic tapes (and probably on optical disks in the not too distant future). Virtually all data are stored in unprocessed form. After expiration of the proprietary period of one year after the date of observation the data (and the calibration observations that go with them) can be requested by outside users. They are not accessible interactively. Requested data will be sent to the requester on a FITS tape.
2. The Observations Catalogue, containing all relevant parameters of, currently, some 100,000 observations, is kept on-line on one of the VAX/VMS computers of the Royal Greenwich Observatory. The Observations Catalogue is available for interactive browsing to outside users. One can select observations by a large number of keys and request their extraction. A subsequent paragraph will describe this process in a summary fashion.
3. The Archive Index of individual packets of information, the top level of which is kept on line. Browsing the index in the same way as the Observations Catalogue can be useful to the ING engineers but is of little use to astronomers.
4. The file of archive users. This file with users names, their addresses, e-mail addresses, *etc.* is used by the archive manager when mailing requested data. Its maintenance is one of the safeguards against violations of the proprietary rights of the original observers.
5. Administrations of transport-tapes and archive volumes, statistics file. Although on-line, these administrations and the statistics are of little significance to the outside user of the archive.

2.2 The WSRT archive

Ever since day one, the observations of the telescope have been properly archived. Virtually all observations with the WSRT are carried out by the observatory staff in a "service" mode; data processing is also done in a controlled environment and is well documented. Off course, the processing as well as the techniques of maintaining an archive administration have been developed considerably over the two decades of the WSRT's existence. For instance, the calibration procedure for each observation was recorded only on paper in the early years. As a result, retrieving the earliest observations and their calibrations from the archive will involve more effort than retrieving the later ones.

In the current procedure for archiving and calibrating, copies of the tapes, produced by the on-line computer of the WSRT, are transported to

NFRA's headquarters in Dwingeloo. All on-line corrections are recorded along with the data in an internal tape format. Before archiving in essentially the same format the data are checked for consistency. Since the beginning of 1990 the archive medium is optical disk. Before that time it was magnetic tape. The archive tapes will all be copied to optical disks in the near future. The archiving process includes building an Observations Catalogue and maintaining an administration of the archive volumes. A special team (the NFRA-WSRT reduction group), is in charge of determining correction parameters from the calibration observations. The tables of correction parameters are stored as part of the archive. The original proposers of the observations receive a copy of the archived data with or without "standard" corrections applied. Archive users have the same options.

Before 1979, the standard data processing was carried out at the Leiden University. As a consequence, the archive of the early observations still resides there. In the next few years it will be transferred to the NFRA headquarters in Dwingeloo to be integrated with the more recent part of the archive.

The organization of the archive of WSRT observations is quite similar to that of the ING observations. The archive is maintained by the WSRT Reduction Group of the Netherlands Foundation for Research in Astronomy, Dwingeloo, the Netherlands. It consists of the following components:

1. Observed and ancillary data, stored in well defined internal WSRT formats on optical disks and on magnetic tapes. Depending on the date of observation the data are either stored in unprocessed form or with standard calibrations applied. In the latter case the corrections can be undone on request. After expiration of the proprietary period of two years after the date of observation the data can be requested by outside users. They are not accessible interactively. Requested data will be sent to the requester on tape. The tape is either in WSRT format or in UV–FITS format usable with NRAO's AIPS package.
2. Calibration information and correction tables. This information is the result of the standard procedure of determining the best possible corrections for instrumental, atmospheric and ionospheric effects. Corrections can be applied on request. Some users prefer not to apply any or all of them because better results can sometimes be obtained by using self-calibration techniques.
3. The Observations Catalogue, containing the most relevant parameters of the observations is kept on-line on the MicroVAX cluster of the Netherlands Foundation for Research in Astronomy. The Catalogue of observations from 1979 onwards (currently some 42,000 entries) is available to outside users for interactive browsing. One can select

observations by a large number of keys and request their extraction. A subsequent paragraph will describe this process. The catalogue of observations made between 1970 and 1979 is not yet available for interactive use but can be requested for inspection.
4. The file of archive users. This file with users names, addresses, e-mail addresses, *etc.* is used by the archive manager when mailing requested data. Its maintenance is one of the safeguards against violations of the proprietary rights of the original observers.
5. Administration of archive volumes, statistics file. Although on-line, this administration and the statistics are of little importance to the outside user of the archive.

12-3 Accessing the La Palma and Westerbork Observations Catalogues

3.1 General

Although the Observations Catalogues of both archives have different formats and contain different keywords, they can both be accessed through the same archive query program, ARCQUERY. Both catalogues, and also the UKIRT and JCMT catalogues, that will be incorporated into the system later, are described by a catalogue description file. Despite the different formats and contents of the catalogues, all of them have in common that searches can be made specifying coordinates in equatorial, galactic or ecliptic systems and by object name. In order to speed up coordinate searches the entries in the Catalogues have been indexed via a set of 6500 cells in equatorial coordinates. The paragraph "Querying Observations Catalogues" will describe the major features of ARCQUERY.

3.2 Access modes

In order to accommodate a large variety of archive users, located at different distances from the sites where the master catalogues of the archives are maintained, ARCQUERY supports three basic access modes:
1. Direct access to the master catalogue. In this mode all commands are executed directly and interactively. Direct access includes remote logins, and local logins with DECnet access to the master catalogue files. In the latter case the initial part of the session may be slow because data transfer via DECnet is inherently slow. All further processing is done on the user's computer. For remote logins both the Royal Greenwich Observatory, Cambridge and the Netherlands Foundation for Research in Astronomy, Dwingeloo have captive accounts

with username ARCQUERY for which no password is required. Obviously, querying the catalogues is the only activity possible from these accounts.

2. Direct access to a local copy of the catalogue, mailed access to the master catalogue. All commands are executed interactively using the local catalogue copy. After making a selection the user has two options at the end of the session:

   - just the requests for data are sent by electronic mail to the master catalogue computer, where they are appended to the archive request queue, or alternatively,

   - the log file of the entire local query session is mailed in order to be re- executed on the master catalogue by an archive-mail-server. The result of this process is e-mailed back to the user. This option is useful especially if the local copy of the catalogue is outdated, because semantic mistakes are more easily avoided than when running a dummy session in mode (3).

3. Mailed access only to the master catalogue. When neither the master catalogue nor a local copy are directly accessible, but the ARCQUERY program is available locally, a dummy query session can be executed. The commands are stored in a log file to be e-mailed to the archive-mail-server to be executed. In this mode syntactic errors are caught by the program, semantic ones are not. In the most basic version of this mode the ARCQUERY commands can be produced by means of an editor. Clearly, this requires some confidence and experience on the part of the user.

Mode (2) requires a copy of one or all Observations Catalogues to be available locally. Modes (2) and (3) involve the availability of a local copy of the catalogue query program (except when an editor is used to produce the commands). Both the Royal Greenwich Observatory and the Netherlands Foundation for Research in Astronomy will be happy to provide copies of the program and/or a recent version of the Observations Catalogues, along with installation procedures, instructions for the system manager and a users manual. Addresses are given at the end of this chapter.

3.3 Querying Observations Catalogues

In order to give the reader some feeling for the philosophy and setup of the catalogue query program its major features are discussed in this paragraph. ARCQUERY is equipped with a context sensitive help facility, which should enable even inexperienced users to find their way about the system. A fairly comprehensive user's guide as well as a Primer to the La Palma archive are available on request.

Once logged into the account ARCQUERY the user will see a menu of functions he or she can perform. The most important of these are:

HELP (or ?) to provide help starting from the top unless one specifies a topic; a tutorial primer will be added soon;
SHOW_CATALOGUES to show the catalogues available for searching along with the access mode to each one of them,
ARCQUERY to enter the catalogue query program,
EXIT to leave the query session.

Before anything can be done the user needs to specify his or her initials. The initials are used to identify all files connected with the query session, including files in which the user can save for a subsequent session his or her personal default settings as well as intermediate results.

After entering the program ARCQUERY one is prompted to specify the catalogue one wants to query or browse. All actions performed by ARCQUERY are controlled by the user's commands. The most important of them are mentioned below; a full list can be obtained via the help facility or by typing a nonsense-command after the ARCQUERY prompt.

SELECT (SE) is ARCQUERY's most basic command. Its syntax resembles a logical expression in FORTRAN with *keywords* and *values* linked with relational operators and logical operators. Most keywords correspond to items in the Catalogue. However some keywords refer to combinations of catalogue items: *e.g.* SEL EQPOSITION=(RA,DEC,R) will select all observations within a circle with a radius R and centered on equatorial co-ordinates RA, DEC.

Like many other commands SELECT produces an output table in the same format as the Catalogue. It will be used as input table by the next command unless the user explicitly specifies a different one. The initial input table is the Catalogue being searched. Although, in principle, the user can specify the observation he/she looks for in one, probably rather complicated SELECT command, one would normally make a coarse selection first on one or two observing parameters (keywords) and narrow the selection down in subsequent SELECT steps.

CALSELECT (CA) is a command designed particularly for the La Palma archive. It selects calibration observations bracketing the regular observations already selected and adds them to the user's final selection.

REQUEST (REQ) is probably the second-most important command. Once the user made a final selection of observations out of the Catalogue he/she REQUESTs the archive manager to actually send the data on a FITS tape or otherwise.

HELPQUERY (H or ?) may be used at any time to obtain help on the action being performed. The help information is organized hierarchically in the same way as the VMS help facility.

SHOW (SH) is a useful command to recall general information, for instance SHOW DEFAULTS will show all current default settings, SH ITEMS will give a description of catalogue items, *etc.*

EXIT (EX) terminates the ARCQUERY session after sending requests to the archive manager, prompting the user to state whether or not he/she wants the session context (tables, files, defaults) saved for later, spooling all print files, and updating the statistics on archive usage.

In addition to the basic commands mentioned above, a number of others are available. The user's manual describes them in more detail.

12–4 User Profiles of the archives

4.1 The La Palma Archive

The archive software was developed in three stages. The first stage consisted of the software to enter data into the La Palma archive and to administer it properly. This stage was completed in 1986. The second stage, comprising the catalogue query software and the programs to process the user requests was completed in 1988. However, the user manual was not published until June 1989 and no serious advertising was done before that time. It is not too surprising that the users caught on rather slowly to the availability of the archive.

The program ARCQUERY monitors its use: statistics are kept of the number of different users, of the number of times they used the program, and how often they used different commands and with what catalogue keywords. These detailed statistics are more useful for decisions on further development of the program than to the readers of this book. The few numbers below may not be fully representative of the use of a well established observatory archive because of the relatively short time the system has been in use.

Currently, summer 1990, the La Palma archive contains some 90,000 observations stored on 1,500 magnetic tapes (most of them 1600 bpi). It comprises the data from two telescopes in use for approximately 5 years and one in use for about a year. The number of registered users is about 50 now, growing steadily by a few every month. There have been over 200 catalogue queries, resulting in 25 requests for data. One or two of these requests were for large numbers of observations: a total of well over one thousand observations have been released to archive users. The number of requests is expected to grow as the proprietary period of more observations of the 4.2-m William Herschel Telescope expires.

4.2 The WSRT archive

The part of the WSRT archive residing in Dwingeloo (observations from 1979 until mid 1990) is stored on approximately 2,500 magnetic tapes and a few optical disks with a capacity of 2 Gbytes each. The Observations Catalogue has over 42,000 entries. It is obvious that the average number of bytes per Westerbork observation is considerably larger than that of a typical La Palma observation. The number of entries in the WSRT Observations Catalogue grows much more slowly than that in the La Palma catalogue.

Hard figures on the use of the archive of Westerbork observations are not available because the archive has only very recently been made accessible through the program ARCQUERY and no user statistics were kept before that time. However, over the two decades that the WSRT archive has existed it has been used fairly regularly and for different purposes. Examples are studies of variability of radiosources and of expansion of supernova remnants for which observations at different epochs existed.

An aperture synthesis radio telescope generally produces more data than the original observers are able or prepared to analyze. Especially at the longer wavelengths, any field observed contains large numbers of background sources, which are often ignored in the analysis process. Obviously these fields are ideal for statistical research on large numbers of sources. It is easy to think of other research possibilities for this archive, including some original ones on data that have never been published for some reason. The relative homogeneity and the availability of well established calibration procedures add to the ease of use of the WSRT archive.

12–5 Future plans

Providing smooth interfaces to some external software packages like the STARLINK general catalogue-management program SCAR (see chapter 18) is part of the agreement between the NFRA and the RGO (to be completed in 1991). Through this interface a number of the statistical and graphical functions of SCAR will be at the disposal of the archive user.

The plan to incorporate the, as yet virtually non-existent, archives of the JCMT submillimeter telescope and the UKIRT infrared telescope into the same system have already been mentioned in the introduction. It is hoped to realize this plan before the end of 1991, if some manmonths at the Royal

Observatory Edinburgh can be made available. The infrastructure of the archive loading and catalogue query software allows incorporation with very little effort.

The archive of Westerbork observations will be transferred to optical disks in the near future. In that process a new, better designed, Observations Catalogue will be built up. The transition from the old to the new catalogue will be transparent to the archive users.

A useful extension, for which no plans have as yet been made, is to incorporate an option in ARCQUERY that will let the user make a selection of observations from one Observations Catalogue and use that selection directly to search for observations of the same objects or fields in another.

Acknowledgements

The Isaac Newton Group of Telescopes at the Observatorio del Roque de los Muchachos del Instituto de Astrofisica de Canarias (Spain) is operated by the Royal Greenwich Observatory on behalf of the Science and Engineering Research Council (United Kingdom) and the Netherlands organization for scientific research (the Netherlands). The Westerbork Synthesis Radio Telescope is operated by the Netherlands Foundation for Research in Astronomy with the financial support of the Netherlands organization for scientific research. The author thanks Dr. Keith P. Tritton for his assistance in collating information about the Isaac Newton Group.

References

[1] van Diepen G. & Raimond E. (1986), Observatorio del Roque de los Muchachos, Isaac Newton Group of telescopes, *Loading and Maintaining the La Palma Archive*, Operator's manual [Version 2], NFRA Note 474, November 1986.

[2] van Diepen G. (1986), Observatorio del Roque de los Muchachos, Isaac Newton Group of telescopes, *Archive Programmer's Guide* [Version 1], NFRA Note 483, April 1986.

[3] Raimond E. & van Diepen G. (1989), Isaac Newton Group, La Palma, User Manual No. XIX, *La Palma Data Archive Users' Guide*, June 1989 [Version 1]; *identical to NFRA Note 537*.

[4] Zuiderwijk E. J. (1989), A primer to the La Palma archive, *ING La Palma Technical Note*, **69**, 1989.

13

Archiving at NRAO's VLA and VLBA Telescopes

Donald C. Wells
National Radio Astronomy Observatory
Charlottesville, VA, USA

All raw data recorded by the Very Large Array (VLA) since it began operation in 1980 have been archived, test observations and calibrations as well as program observations. The reels of tape are kept in a vault in the control building of the VLA on the Plains of San Augustin in New Mexico; the vault was filled several years ago and so recordings for recent years are in a trailer parked outside the control building. The data are in the public domain, by policy, with a proprietary period of 18 months for program observations. A subset of the observation catalog (all non-calibrator observations with duration longer than one hour) is distributed as a set of floppy disks (MS-DOS format) with a search program (for MS-DOS); the full catalog is available on a computer at the Array Operations Center in Socorro, New Mexico. One technical problem is that the existing catalog doesn't tabulate which reel of tape in the vault has which day's data. Reel numbers must be obtained from a log book at the VLA site. But the data tapes can be located in the vault, they are coded in a modest set of reasonably self-documenting formats, and a portable program which is exported by the National Radio Astronomy Observatory (NRAO) in all AIPS (Astronomical Image Processing System) kits can be used to read files copied from them.

The Very Long Baseline Array (VLBA) is under construction, and will be operational in 1992. All data produced by the VLBA's signal processing complex (the *Correlator*) will be archived. Data ownership and proprietary period policies will identical to those of the VLA. The architecture for the VLBA archive has been specified (see sections 2.7, 2.8 and 4.3.10 of [1]) and construction funding is assured; the construction has been deferred until 1991 in order to maximize options for choice of media (4 mm and 8 mm magnetic cartridges are currently the leading candidates). The plan specifies that the VLBA operators will be available to mount and dismount archive media on request. The Correlator includes a relational database system to maintain the catalog for the archive. The Correlator control computer and its DBMS will be attached to the Internet, with appropriate query tools provided. Distribution of archive data over the Internet will be supported; indeed, it is likely that network transmission will be the most common means of data transfer even inside the Array Operations Center by 1995. Distribution by various popular magnetic media will also be supported.

An important feature of the VLA and VLBA archives is that original observers have essentially no special advantages over archive researchers: they get the same data in the same format. This is an essential policy and technical principle for good archive design.

At the present time (1990) the VLA archive retrieval process is tedious because it is not really funded: there are no operators or programmers assigned to assist archive researchers and there is no network access. The funding problem also prevents critical maintenance, such as periodic recopying of the media. It is known that some of the data are no longer easily readable, if at all, and the degradation of the signals continues steadily. Considerable resources are needed in order to assign people to re-record the data on new higher density media, to operate the retrieval process, and to improve the catalog and make it visible through the networks. Unfortunately, at present, no funding strategy has been found and a continued VLA archive operation is uncertain.

The VLBA construction funding will not support the long-term operational requirements of its archive maintenance, of course, and therefore the VLBA will probably suffer from many of the same operational limitations as the VLA unless some new source of funding is found. I conclude that for a ground-based archive policy to be fully successful over a long period of time (> 10 years) it will be necessary to fund it as a separate budget item, distinct from normal operating budgets.

References

[1] *Software Architecture for the VLBA Correlator* (1989), VLBA Correlator Memo No. 95, 29 September 1989.

14

ESIS
A Science Information System

Miguel A. Albrecht

ESA/ESRIN
Via Galileo Galilei
00044 Frascati, Italy

14–1 Introduction

This chapter describes the work being currently undertaken to define and develop the European Space Information System (ESIS). We will concentrate here on the user functionality aspects of the project, and refer the reader, for further detail, to Albrecht *et al.*, 1988.

ESIS aims at providing the Space Science community with a set of tools to access, exchange and handle information from a great variety of sources including space mission archives, databases of scientific results, bibliographical references, *white–* and *yellow–page* directories, and other information services. Access to all information contained in ESIS shall be possible through a homogeneous interrogation and data managing language. This language will support queries formulated in *discipline oriented* rather than *computer oriented* terms. A *personal database* environment will support the retrieval of data from different sources into a common homogeneous format: this constitutes the basis for cross correlations and other further processing.

A Pilot Project running in the timeframe 1988–1992 will demonstrate the validity of the concepts adopted and will establish a basic infrastructure for further developments.

14-2 User needs

The general requirements of space scientists on ESIS can be summarized in the points listed below.

(a) Open existing space data holdings to non-experts.

Due to the high level of specialisation required to access current archiving facilities, data is being utilized only by small 'mission-specific' user communities. A global reference database could provide non-experts with an overview of available data. This requirement calls for an integrated data view model and a layered access structure.

(b) Integrate past and future mission archives.

Existing and future archiving facilities should be integrated into the system without losing their identity and/or independence. For past missions this means that current investments are preserved and that local user communities are not concerned by the integration. For future missions, independence from the system means the flexibility to set up their facilities taking advantage of state-of-the-art technology without making compromises. This requirement calls for an open architecture.

(c) Integrate data retrieval with information exchange.

Finding out what data sets are of relevance for a particular research project, retrieving these data sets, processing them and possibly exchanging them with other collaborators should all be activities integrated into one single environment. Provision should also be made to interface in a transparent way other information systems and telecommunication networks. This requirement calls for a uniform access method to all services available through ESIS.

14-3 The pilot project

In order to test out feasible concepts both on the system engineering as well as user interface sides, a pilot project has been set up to run over the period 1988–1992. During this period the system functionality and its architecture is defined and tested in a 'real-life' environment, *i.e.* it is actually used in day-to-day scientific work. It is assumed that a science information system can be designed independently of the science domains that it serves: being the discipline specific aspects confined into well defined, exchangeable layers. In fact, ESIS is meant to serve the whole range of space science domains, beginning in the pilot phase with Astronomy & Astrophysics and Space Physics but expected to include Earth Observation and Microgravity sciences later on. At the time this article is written, the pilot phase has gone half way through. The system architecture has been designed, a first set of user functions have been defined and the system is expected to enter the implementation phase in the first months of 1991.

3.1 Pilot Project configuration

The elements that form the complete system will be briefly described below. Due to resource limitations both in manpower and infrastructure, a number of decisions had to be made in the configuration of the *baseline* system. The number of databases to include, the number of user hardware platforms, the associated network protocols to support, *etc.*, all define the starting system, but great care has been taken to isolate dependencies in order to ensure expansibility and further development.

3.1.1 Databases included in the pilot project. The archiving facilities and databases to be included in the pilot phase are (most of these systems are described in other chapters in this book):

- the IUE Vilspa DB, in Villafranca, Spain;
- the EXOSAT Database, at ESTEC, Noordwijk, the Netherlands;
- the SIMBAD Database, at the Centre de Données astronomiques de Strasbourg (CDS), France;
- the Space Telescope archive, at the ST–ECF, Garching, Fed. Rep. of Germany;
- the World Data Centre C1 and the Geophysical Data Facility managed by the Rutherford Appleton Laboratory, Chilton, Didcot, United Kingdom;
- relevant bibliographical databases offered on-line by the Information Retrieval Service (IRS) at ESRIN, Frascati, Italy.

3.1.2 Information services. Value added information services will also be included. They comprise the 'classical' set of telecommunication functions like electronic mail, file transfer and management, remote login, as well as group communication tools like electronic conferencing, bulletin boards, *etc.* A directory service for users, institutions, instrumentation and software packages will also be established that will allow both *white–* and *yellow–page* searches.

3.1.3 The Backbone network. Since ESIS has to connect distributed, heterogeneous data systems all over Europe, the underlaying telecommunication infrastructure plays a key role. The ESIS backbone network has been established making use of the telecommunication facilities provided by ESANET. In addition to the telecommunication lines, two kinds of elements form the backbone network: the ESIS access points and the service front-ends (also called *service shells*). A number of Vax computers have been procured to host these elements covering today six sites, and it is expected that within 1992 a series of further procurements will complement this infrastructure to cover all ESA member states.

14–4 System architecture

ESIS is defined through a layered architecture that confines the system functions into the following three layers: the Pilot Distributed System (PDS, the bottom layer), the Query Environment and the Correlation Environment (see figure 14–1). Each layer requests services from the one below via a set of interfaces. All functions, regardless of their location, both architectural and geographical, are accessed via a unique environment: the *User Shell*.

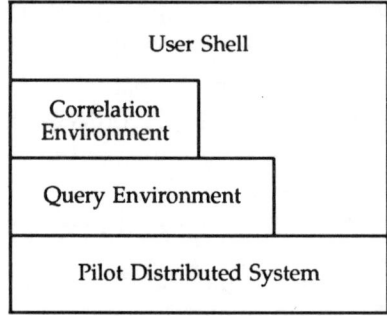

Figure 14–1: A user's perspective of the ESIS environments

The ESIS User Shell is a software package to be distributed to end users for installation on their local computing infrastructure. The user interface is designed to support a large number of devices including terminals, workstations and personal computers, and offers a selection of dialogue modes such as menus, commands and forms. A user activating the User Shell will reach all ESIS functions through a uniform view of the system, independent of the local computing environment and of the networking method of accessing ESIS. On-line help and tutorial services will be provided locally, in addition to any similar services offered by the data facility, allowing for a quicker response and avoiding unecessary network load.

The Pilot Distributed System will provide information services such as electronic mail, bulletin boards, file transfer, *etc.* as well as transparent message passing services for ESIS applications. Also, the ESIS PDS is concerned with providing controlled and secure access to archiving facilities, and more generally handles secure message passing between generic clients and servers.

The ESIS Query and Correlation Environments will provide information retrieval, handling and storage functions, as well as the possibility to cross correlate data from different sources. The user functionality defined so far within these two environments will be described in better detail below.

14–5 Science Oriented Information Retrieval

A key issue in ESIS's capability to integrate heterogeneous information sources is the definition of a *unified data model*. In this section we will describe how such a model has been built, what are its key components, and what user functions will eventually make ESIS a science information system.

5.1 The heterogeneity factor

In order to understand how a unified data model can be built, it is first necessary to briefly summarize the levels of heterogeneity that are to be overcome. Heterogeneity is reflected at many levels:

PHYSICAL HETEROGENEITY
: Data stored by different database management systems (DBMS) or file systems on a variety of different hardware platforms generally have different file formats, index mechanisms, record lengths and so on.

SYNTACTIC HETEROGENEITY
: The different catalogues and archives have their own denomination of objects, their different spelling rules, different abbreviations, different fault tolerance and, of course, their own query language. The result is that it is practically impossible for a novice user to remember exactly how to syntactically formulate his database queries.
: Also different (mis)spelling causes problems. In the INSPEC database of bibliographical references, for instance, 'The Monthly Notices of the Royal Astronomical Society of the UK' is spelled in (minimum) 10 different ways. In general, spelling mistakes resulting in *uncontrolled* keywords, may lead to loss of information.

SEMANTIC HETEROGENEITY
: On the semantical level, heterogeneity appears when different *meaning* is attached to one single term (homonyms), or inversely when different terms express the same meaning (synonyms).

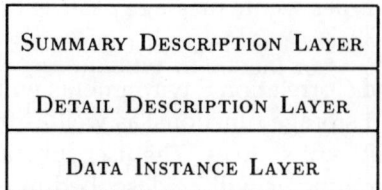

Figure 14–2: ESIS Query Environment Information Access Layers

CONTEXTUAL HETEROGENEITY
Contextual heterogeneity primarily refers to the fact that equal terms can be interpreted differently in different science disciplines.

5.2 A layered data view model

Generally speaking, connecting or combining information from different sources implies providing an overview of the available information, transparent to *where* and *how* it is stored. On the other hand one surely wants to have access to the data records themselves. These requirements call for a structure of layers that provide different levels of abstraction. In fact, this is how ESIS overcomes all levels of data heterogeneity, the upper layer provides a unified view while the lower layers contain archive specific detail information.

SUMMARY DESCRIPTION LAYER
The summary description layer contains high-level key information about all accessible information in a uniform, homogenised way. The conceptual schema for this layer will be drawn from the union of all semantic definitions for the disciplines involved in ESIS. This key information should be suitable for making an initial determination of the potential usefulness of an information object (*e.g.* a data granule). In this layer all heterogeneity levels are overcome except contextual heterogeneity, for there will be one of these layers for each discipline.

DETAIL DESCRIPTION LAYER
This is primarily what the ESIS pilot project databases offer today. It is still information about something, but at a more detailed level than above.
There may be different kinds of detail descriptions:

- *Database catalogue entries* (*e.g.* an observation log), a detailed description of lower-level data available at that data centre.

- *Quick-look* products, *i.e.* data sets of limited size and resolution created to provide an understanding of the type and quality of available full resolution data, and/or to enable the selection of intervals for the analysis of physical events.

- *Examples*, simply limited-in-size examples of data instances, which could give a clue to the kind of data in question. Instead of, for instance retrieving a full data set, the user can request an example, in which case the system could provide 1–2 pages of the full data instance.

- *Abstracts*, a textual summary of the contents of the actual information item. These could be descriptions of publications, quantitative results, catalogues, databases, data systems, *etc*.

DATA INSTANCES LAYER

At this layer we are confronted with the actual data themselves (and not descriptions thereof). In the case of observations this may be an image, a spectrum, *etc*., while in the case of publications, it could be a hardcopy of the actual article.

Naturally, information from the last two layers is specific to each database, and so of heterogeneous nature. However, when retrieved over ESIS, physical formats will be converted to an internal common standard allowing further manipulation at the physical level.

5.3 The ESIS Reference Directory

The summary description layer mentioned above reflects the unified data model for a particular discipline. In order to model the way scientists from different science domains see their *reality of interest*, semantic networks have been designed that constitute the basis for the eventual retrieval system. Since each discipline has its own semantic denominations, one such network has been worked out for each of the disciplines that ESIS will serve: Astronomy & Astrophysics, and Space Physics. The design of these schemata has been conducted by the Centre de Données astronomiques de Strasbourg (CDS) for astronomy and by the Rutherford Appleton Laboratory (RAL) in collaboration with the Sheffield University for space physics, under contract for ESA, and has involved the participation of users through workshops and working groups (Egret *et al*., 1990). Figure 14–3 shows the latest available version of the astronomical semantic network drawn using the Entity–Relationship diagramming technique. Boxes represent entities, diamonds represent relationships among entities; attributes for both are given when defined. A query is composed by selecting entities and/or relationships possibly qualifying them by giving values to some attributes.

In order to be able to answer user queries, the retrieval system must take into consideration both the user data models, as defined by the discipline semantic networks and the union of all local archive data descriptions. The database in ESIS that integrates these two aspects is called the *Reference*

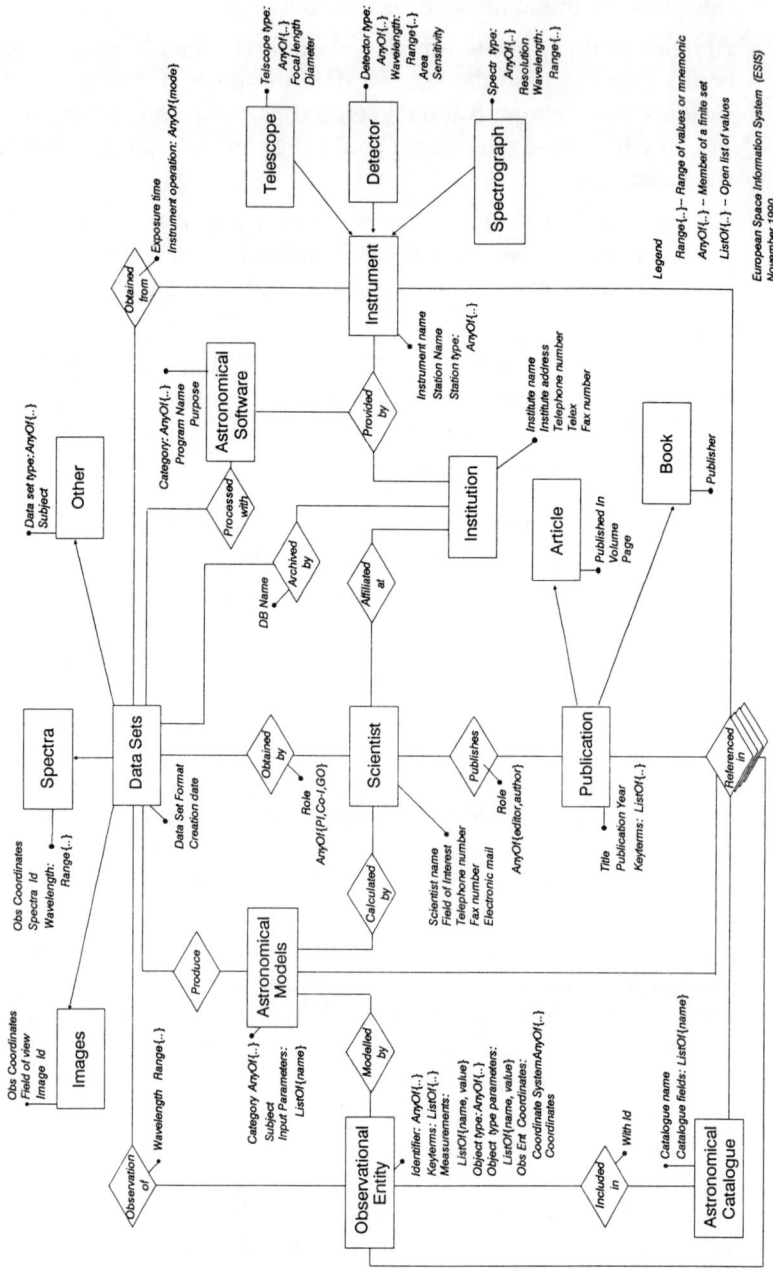

Figure 14-3: ESIS Astronomical Query Language: Semantic Definitions (November 1990)

Directory and it includes both the global data dictionary as well as the summary description layer. The Reference Directory will play a fundamental rôle in the system's ability to provide quick responses to user requests.

The key difference between the Reference Directory and 'classic' approaches (see *e.g.* Larson, 1989) solving the problem of heterogeneous, distributed databases, is that in ESIS heterogeneity is overcome not only at the level of the data description, but also at the level of their *semantic* contents (see section 5.1 above). Another distinctive feature of the Reference Directory is that the system can take full advantage of the possibility to resolve queries at a *summary information* level, *i.e.* needing only a small subset of the global data contents available in the system. After the user has gained a clear idea of the datasets of interest to his/her needs through inspecting summary descriptions thereof, a highly efficient fetching method can be applied to actually retrieve the datasets in question. In fact, not only datasets but also detailed descriptions, quick-looks or other information items can be retrieved in the same way. The method used here is that of 'intelligent pointers' that are embedded in the Reference Directory for each datum described therein. These pointers can be thought of objects that are activated upon request and that contain enough knowledge to perform an information retrieval operation. The details of this operation are concealed within the object's interface: they could be small SQL programs or query language constructs, they could also be arithmetic operations, requests to a server on the network or simply file fetching commands.

5.4 Discipline specific query languages

The superset of functions available to the user can be described in terms of what can be called a *Discipline Query Language*. In fact, it has resulted from the definition of the system that each discipline has not only its own semantic definitions but also specific query primitives, *e.g.* {select-by-cone} in astronomy.

In general we envisage that, through the Query and Correlation Environments, users will gain access to the global information contents of ESIS as well as access to personal databases. Personal databases could contain data imported from external sources or data retrieved from the global database. Also the exchange of databases with other users will be supported.

The user will interact with the environment through a suite of so-called *interaction paradigms*. These paradigms will support command or query input in a variety of ways and will display results in a form suitable to the data. Database *views*, *i.e.* the capability to tailor the amount of information the user sees from the database, will play an important rôle. Personal *views* of databases will allow a customized access to the information stored. Globally pre-defined *views* will allow groups of individuals to share the same way of accessing data. The formulation of queries will support the combination of any of the interaction paradigms with any user defined view. The overall environment will be enhanced by including standard support tools and discipline specific ones such as coordinate-system transformations and physical-unit conversions.

Among the functions that will be available through the User Shell the following should be highlighted:

QUERY AND COMMAND FORMULATION

Command or query formulation will be supported through a set of interaction paradigms that will be used in any combination at any given time. Depending on the kind of input required, the most suitable method will be selected either by the user himself or by the application as the default setting. The interaction paradigms mentioned above will be complemented by discipline specific query primitives, such as *search by {cone —ellipse — polygon —..}*.

OBJECT (DATA) VISUALIZATION ACCORDING TO STRUCTURE

The display of information items will be driven by their internal structure; in this way a vector plot will be visualized as graphs while a database record could be displayed in a form. Database structure will be shown as entity–relationship diagrams by default. Using an object–oriented terminology, we can say that sending the *message* {display-yourself} to any information item, will produce a visualization of the object's contents in the form best suited to the object's structure.

PERSONAL VIEWS AND DATABASES

A basic requirement in science information system is the ability to manipulate data according to the needs of a research project. The basis for such a capability is implemented in ESIS both through functions to keep and manage personal databases and functions to look at the database contents in particular ways (views) that are created circumstantially. A graphical concept manipulation language (*see* Catarci & Santucci, 1988) will allow users to create and manage databases and views in a user friendly way. Entity-Relationship diagrams will be used as input paradigm for all functions related to the manipulation of database structure, *i.e.* view manipulation and management of personal databases.

The personal database capability will form the basis for multi-mission, multi-spectral analysis. It will allow users to retrieve data from different sources, that, when delivered will share the same physical, syntactic and semantic format, paving the way towards cross correlations accross spectral domains.

SUPPORT TOOLS

It is envisaged to provide within the system general support tools for information manipulation. These functions are not yet fully defined; however, at the time of writing, they will include a 'scientific spreadsheet' capability, an astronomical calculator and means to support user notes, glossaries and lexica. These tools will support coordinate system and physical unit transformations, the handling of user catalogues and their manipulation, *etc.*

INFORMATION EXCHANGE

The possibility to export data from the ESIS environment onto external applications as well as the complementary import function will enhance the overall functionality of ESIS. A set of standard external formats will be supported such as the formats in use by the astronomical community for spectra and images (FITS) and formats for graphical metafiles. Further, fully integrated support functions like mail and data set transfer will allow the transparent exchange of data sets, personal databases, and other processed information.

14–6 Conclusions

At the time of writing (November 1990) the ESIS pilot project has gone half way through. An overall system structure has been defined, leading to a layered, open architecture. A science oriented information retrieval system has been designed that takes into account all levels of heterogeneity among the different databases, and solves the integration problem by defining a homogeneous layer containing summary descriptions of the data. The conceptual schema of this layer is given by semantic definitions, that, for each discipline involved in ESIS, have been worked out, under ESA contract, by centres of excellence with end–user participation. Discipline specific query functions enhance the system interrogation capabilities. Personal databases will support multi-spectral analysis. Support tools will enhance the overall functionality to make ESIS a Science Information System.

Further Information

For further information and to receive the ESIS Newsletter and workshop announcements please contact:

The ESIS Project
ESA/ESRIN
Via Galileo Galilei
I-00044 Frascati, Italy
Tel: +39 (6) 941801, Fax: +39 (6) 94180361
ESIS::ESIS on SPAN
ESIS@IFRESA51 on EARN

Acknowledgments

The author wishes to acknowledge the enthusiasm and support experienced from all individuals that have actively participated in the various phases of the project, in particular, from the ESIS teams at ESRIN, at the CDS, at RAL and at Sheffield University. All members of the ESIS Steering Committee as well as of the Astronomical Databases and Space Physics Working Groups have injected substantial momentum into the project, their contribution has been highly appreciated.

References

[1] Albrecht M.A., Russo G., Richmond A., Hapgood M.A. (1988), "Towards a European Space Information System, Vol. 1, General Review of User Requirements", *ESA internal report.*
[2] Albrecht M.A. & Bodini A. (1988), "Towards a European Space Information System, Vol. 2, Implementation Plan", *ESA Internal Report.*
[3] Catarci C. & Santucci G. (1988), "Query by Diagram: A graphical query system", *Proc. 7th Int. Conf. on ER Approach,* North Holland, Amsterdam.
[4] Larson J.A. (1989), "Four Reference Architectures for Distributed Database Management Systems", Computer Standards & Interfaces **8**, 209.
[5] Egret D., Ansari S.G., Denizman L., Preite-Martinez A. (1990), The ESIS Query Environment: Final Report, *ESA Internal Report.*

15

The NASA Astrophysics Data System

James R. Weiss & John C. Good
Jet Propulsion Laboratory
Infrared Processing and Analysis Center
California Institute of Technology
MS 100-22
Pasadena, CA 91125, USA

15–1 Introduction

The goal of the Astrophysics Data System (ADS) Program, sponsored by the Astrophysics Division of NASA, is to provide astronomers with easy access to astrophysical data and analysis capabilities without regard to the location or sub–discipline orientation of the user, the data, or the processing tools involved.

The specific near–term objectives of this Program are the following: to interconnect Astrophysics users, the data, and processing elements of the ADS; to provide efficient and effective tools for locating remote multi-spectral Astrophysics data; to provide a uniform user interface for basic data queries; and to generally facilitate the development of information system services devoted to the acquisition, processing, analysis, and management of Astrophysics data.

In response to these objectives NASA has developed an information system infrastructure which transcends system component heterogeneity and provides the basic required services for building data analysis systems within Astrophysics. The ADS is currently a distributed, multispectral, service–oriented information system, which networks the existing and planned Astrophysics data centers and archives (referred to as 'nodes') with the user via the NASA Science Internet (NSI).

Besides basic infrastructure, the initial implementation of the ADS provides User Services for location, retrieval, and some analysis of on–line data from space-based NASA missions, and for retrieval of documentation relevant to the data and to astrophysical research in general. Taken together, these are loosely referred to as 'Directory Services', and are described below. Many more User Services are expected in the near future.

In its initial release, the ADS will consist of the following six physically distributed primary nodes (see section 4.3 for responsibilities of primary nodes): Harvard Smithsonian Astrophysical Observatory (SAO), the Caltech Infrared Processing and Analysis Center (IPAC), the Goddard Space Flight Center (GSFC) and University of Colorado (CU) International Ultraviolet Explorer (IUE) facilities, the Hubble Space Telescope Science Institute (STScI), and the GSFC National Space Science Data Center (NSSDC).

15-2 System Development

2.1 Design Philosophy

The Astrophysics Data System was designed to deal effectively with the data volumes and analysis needs that will be encountered during the era of the Great Observatories, *i.e.* for the next few decades. During such a long period of time, computer hardware and software can be expected to evolve through several generations of technology and design philosophies. The ADS therefore required an approach which was independent of today's specific hardware, software and configuration choices, and provided an infrastructure likely to be evolvable over this whole era.

To accommodate the anticipated growth and changes in both technology and requirements, the ADS employs a client/server architecture which allows services and data to be added or replaced without having to change the basic architecture or interfaces. The ADS design is modular and layered, which enables smooth evolution of the hardware and software without interruption of service to the user community. In addition, the ADS will provide all on-line information necessary to use the ADS services and data; and its design enables remote updating of the software services.

Conceptually, the software can be divided into two categories. Core services are those which provide the basic system functionality upon which user applications or User Services can be built. User Services will provide science support and analysis functions required by the investigator. In its initial release, the ADS User Services consist primarily of Directory Services which provide information on data holdings within the NASA Astrophysics archives and, in most cases, to access the data itself.

There are two guiding philosophies for ADS development. The first is that the ADS Project itself should concentrate on the enabling technology and organizational requirements of the system, while data and most applications services remain the responsibility of projects, such as those mentioned as primary nodes, which have data archive and dissemination as part of their charter.

The second ADS principle is that services and data should be placed where they will be the most effective. For data, this has meant co-locating the data with the human expertise that understands it best and is therefore best able to analyze it, correct it, and explain it to the community. For software, this has meant co-locating the user interface with the user, database operations with the data, and processing with existing processing power.

2.2 Implementation approach

The ADS development has been a collaborative effort, which intimately involved science users from NASA's research community who work in close collaboration with data system designers, developers, operators, technologists and managers from NASA's Science Operations community and innovative elements from the commercial sector. This diverse ensemble is organized into a tightly-knit working group, providing in one place the breadth of perspective needed for this challenging task.

The implementation approach used by ADS has been one of collaborative testbedding directed by (and involving) the entire working group. This approach uses rapid prototyping to validate assumed requirements and develop design alternatives as well as to verify ultimate design concepts. With it, the applicability of key selected technologies was verified quickly, utilizing the vast experience of the Astrophysics user community, NASA Centers, and industrial affiliations.

15–3 System Design

The ADS infrastructure provides a highly distributed user community with data access and functionality from a distributed, multispectral, heterogeneous set of data systems. The basic elements of this infrastructure are the networks, core services, standards and protocols described in the following sections.

3.1 Networks

The NASA Science Internet (NSI) provides the network backbone and tail circuits for the ADS. NSI consists of the Space Physics Analysis Network (SPAN), based on the DECnet protocol, and the NASA Science Network

(NSN), based on the TCP/IP protocol. NSN encompasses the Program Support Communications Network (PSCN) and NSFnet services. The first release of the ADS will support TCP/IP; however, it is intended that the ADS will eventually support the DECNET protocol. It is the intention of the NSI Program to provide a smooth transition from the current set of protocols to International Standards Organization (ISO) protocols as they become available.

3.2 Core services

The operation of the ADS involves servers, services, clients, and transactions. Servers are the physical computers where the work is done. Services are the programs that do the work, whether it be database access, interprocess communications, or user interface. A client is a process which requests services (and itself may be a service for someone else). A transaction is the coordinated interaction with multiple datasets by a group of processes in response to a single request from a client.

The mapping of services to servers in the ADS is unconstrained: some servers simultaneously provide several simple services while some complex services are segmented across several servers. ADS services need not be confined to a single node; in fact, all core services run at all nodes. Similarly, the number and kind of services that can be mounted on the ADS are also unconstrained. Among these services, there are six that are essential in that without them the mounting of the other services would be difficult or impossible. These services comprise, collectively, the Core Services of the ADS: the User Agent, the Message Passing Service, the Transaction Manager, the Data Independent Access Method, the SQL Servers, and the Factor Space Access Method.

3.3 User Agent

The User Agent (commonly referred to as the user interface) is the service that provides the casual end user's local working environment. Through it he can locate data (descriptions and catalogs) of interest, access datasets either individually or in combination, and select from among them one or more subsets that he wishes to import into his local working environment for further analysis. None of this requires that the user knows which servers execute the services he invokes.

The User Agent provides the combined functionality of: (a) a full-featured text management facility that supports browsing, plain-text inquiry and cut-paste editing of selected files; (b) a first-order complete relational database facility including support for Structured Query Language (SQL) and Query by Example (QBE); (c) a context-sensitive help and dynamic tutorial facility; and (d) custom menu generation and scripting facilities.

In the initial ADS release, the Knowledge Dictionary System™ developed by Ellery Systems Inc. is being used to implement the functionality of the User Agent.

3.4 Message Passing Service (MPS)

The Message Passing Service provides remote inter-operability among other services. To a casual end-user interacting with the User Agent, the MPS is invisible. By contrast, the MPS provides to the application programmer (and, of course, to the ADS system programmers) what the User Agent provides for the casual end-user: an environment in which to access the services of the ADS, without knowing which servers will execute the programs or the services these programs will themselves ultimately access.

At a minimum, the MPS implements the functionality of: (a) Remote Procedure Call (RPC), a mechanism by which the subroutines of a program can call each other as if they were executing on a single server when, in fact, they may be segmented across several servers; (b) Remote Process Invocation (RPI), a mechanism by which a program on one server can spawn programs on other servers; and (c) Remote Inter-Process Communication (RIPC), a mechanism whereby two or more programs running concurrently on different servers can communicate among themselves. In the initial release, the ADS Message Passing Service conforms to the Advanced Network System Architecture (ANSA).

3.5 Transaction Manager (TM)

In a distributed system, data transaction integrity becomes a more difficult issue than on a single CPU. A process running on one system and accessing either data or processing on another system has to both know when the other resource disappears (*e.g.* the other machine goes down) and how to recover from the loss.

The service in the ADS that provides process and resource management is called the Transaction Manager. The TM provides one basic function: ensuring, without further client intervention, that all resources involved in a transaction are properly synchronized and released regardless of the destiny of the transaction (success or failure). In other words, the TM (if used) ensures that all the data involved in a transaction reach the final state (the transaction is successful) or that all of it is returned to the initial state (a blanket reset).

The TM is both an optional and a passive service: optional in that it must be explicitly invoked by the client; and passive in that, as a peer-to-peer system, there is no mechanism by which the TM protocol can be enforced, and the only programs that are guaranteed to participate in the TM protocol are the core services of the ADS. To encourage the use of the TM by end-users and application programmers, the protocol has been kept as simple as possible, requiring only four commands: Begin, Lock, Upgrade, and End.

Begin is a signal that a process is starting and returns a unique Transaction ID that the process must use in all subsequent calls to the TM. Lock signals the intent of the process to access the resource (*e.g.*, a record in a file) named in the call. Upgrade signals the intent of the process to modify a previously Locked resource. And End signals that the process is terminating and the mode of termination (success or failure). For the initial ADS release, the procedures that implement the Transaction Manager as an integral part of the Knowledge Dictionary System™ will be exported as discrete services through the Message Passing Service and will be accessible by Remote Procedure Calls (RPCs).

3.6 Data Independent Access Method (DIAM)

The service that provides inter-operability among the various Database Management Systems (DBMS) in the ADS is the Data Independent Access Method. The DIAM is required in the near-term releases of the ADS in order to provide the user with a uniform relational view of all the catalogs, even though some of these catalogs are still maintained using DBMSs that do not yet support the relational SQL standard.

Thus, the functionality of the DIAM is: (a) to accept SQL statements generated by the User Agent (either directly or indirectly using the QBE translator); (b) to convert them to whatever language is supported by the DBMS hosting the catalog(s); (c) to invoke a remote process to submit the query to the DBMS for processing and to collect the resulting table(s); and (d) to return those results to the User Agent (or other client) that issued the request. In the initial ADS release, the Distributed Access View Integrated Database (DAVID) system is being used to implement the functionality of the DIAM.

3.7 SQL Servers

For systems that do support SQL programming interfaces there is a more efficient, direct-access data interface. This consists of a set of SQL Servers, each one tailored to the DBMS being accessed but all using the same RPC protocol when talking to clients. The net effect is that clients can query widely varying databases using a standard query syntax (SQL) and program interface (RPC). This functionality can fit under the DIAM described above as a special (but very common) case or can be accessed directly by clients that construct their own SQL queries.

The basic protocol presented by an SQL Server to potential clients involves a simple dialog as follows: after successfully connecting to the service, the client submits an SQL statement (straight text in standard SQL syntax). The service responds with an acknowledgment of receipt. The client (not being blocked) can do other things or can immediately start polling the status of the request. The service responds in the negative if it is still processing or in the affirmative if the request has been completed. The client then asks for all or part of the data and the service starts sending it.

All the client–generated messages (SQL statement, request for status, request for data) are in a standard, predefined syntax. The same is true for the service–generated ones (query acknowledgement, status report, data). All SQL Servers that adhere to this syntax can be used interchangeably, although they may, in fact, have widely varying implementations.

3.8 Factor Space Access Method (FSAM)

The service that provides uniform subject matter indexing across all the textual data in the ADS (by far the greater part of which is not stored in any structured database) is the Factor Space Access Method™. The Factor Space is an n–dimensional euclidean space, the axes of which are statistically constructed to account for the variance and covariance in expert judgments made by Astrophysicists about the relevance of a piece of documentation to different subject matter contexts.

The functions of the Factor Space Access Method service effort are the following:

1. to scan all documents available through the ADS and compute their subject matter profiles as a sequence of one or more vectors in the Factor Space;
2. to similarly analyze natural language requests and to search the Factor Space for relevant documents;
3. to monitor the distribution of vectors in Factor Space for clusters (*i.e.* undifferentiated document groupings);
4. to periodically generate, disseminate, collect and synthesize questionnaires to obtain additional relevance judgments needed to increase the data resolution; and

5. to factor–analyze these additional relevance judgments and modify the number and/or orientation of axes in the Factor Space accordingly.

Functions (3) through (5) also support the generation of new Factor Spaces, either to accommodate new subject matter or to accommodate personalized perspectives on existing subject matter. For the initial ADS release, the procedures that implement the Factor Space Access Method as an integral part of the Knowledge Dictionary System will be exported as discrete services through the Message Passing Service through which they will be accessible by Remote Procedure Calls.

15–4 System Operations

The ADS is a geographically distributed system which is organized, for organizational purposes, into a hierarchical configuration of nodes. The ADS will be composed of Primary and Secondary Nodes, an Administrative Node, and User Sites.

Each node has a portion of the ADS data (often redundantly) and operates as an autonomous element of the system with total responsibility for its own data holdings. The over–all system integrity is maintained through operational and administrative requirements dictated by the ADS Project and enforced by the system administrator. Elements of this responsibility are covered in the following sections.

4.1 ADS System Administration

The ADS will maintain an administrative node whose primary functions are to monitor the system (user/service database, network throughput and connectivity, usage patterns, service availability, and security) and maintain cognizance over core services, documentation, new node integration, and service augmentation. The ADS administrative node will also serve as the top–level coordinator for dealing with user questions, problems, training, and system development, though initial contact responsibility will rest with the nodes.

Configuration control will be the responsibility of ADS Project management with advice from the ADS Working Group and implementation coordinated by the administrative node. Individual nodes will be required to provide for adequate change control (review, testing, notification, *etc.*) for their services. This change control plan will be reviewed for consistency with Program change control procedures when new nodes are proposed.

4.2 ADS Node Administration

Individual nodes are required to provide user support services for their local users and for the remote users of those services for which they are the primary provider. They are also required to maintain configuration files, usage summaries, and other information needed for overall system monitoring.

Each node will provide personnel to act as user consultants, to maintain the software, hardware and databases, and to act as node administrators.

4.3 Responsibilities of Primary Nodes

An ADS primary node is a facility which assumes primary responsibility for the provision of a unique service through the ADS infrastructure. This service can involve the provision of basic mission–specific data, as would be the case for a mission science support center like IUE, or for the provision of an operational service for remote access such as the IPAC-supported NASA Extragalactic Database (NED, see chapter 10 of this book).

All primary nodes participate in the ADS Directory Services by making appropriate databases accessible through the ADS DIAM and/or SQL server interfaces and by providing appropriate documentation describing their supported missions, facilities and data holdings.

Primary nodes must augment their user support staff to provide the adequate level of expertise needed by the user community to interpret the on–line data. They must also provide the manpower required to maintain the data services they uniquely provide via the ADS.

Other User Services beyond the Directory Services may be offered by primary nodes, such as access to specialized processors, operational services, or unique software tools. These services must be supported with appropriate documentation and manpower if they are to be made available through the ADS.

4.4 User/Process Authentication

In order to serve the community effectively, while at the same time maintaining security and proprietary rights to some data sets, both users and services must be registered. The ADS plans to use the KERBEROS software developed by Project Athena at MIT for this service.

In the initial release of the ADS, only user authentication is performed (essentially a login/password step). This step is performed by the User Agent software, accessing the KERBEROS database which is itself replicated but ultimately the responsibility of the administrative node. As the ADS matures it is anticipated that increased usage will drive enhanced requirements for security and user priorities. In the near future, the full KERBEROS capability will be activated, making it possible to authorize users (and services) to use specific services.

Node administration will be expected to collect the data required for this system; since primary nodes are responsible for their own user base, they are also required to acquire and maintain the relevant information on their users and their users' access to all system services. For each of its users, the node will determine which local services that user may access and negotiate with other external service providers for remote user authorization. This information is entered into the database by the node. Each node is expected to support a certain number of local and remote users at its own discretion. Off–line cross–checking and certification is done by the System Administration node. Any services for which the primary node is the principal provider will also be registered in this database and the node will coordinate with secondary providers of that same service for uniform support and user authorization.

4.5 Node Maintenance

Each node will maintain its server platforms, network access hardware, unique ADS service software and resident documentation. Maintenance of the core system software at NASA ADS nodes and sites will be provided through the ADS Project via a contract with Ellery Systems, Inc. Updates to the core system software at ADS nodes and sites will be issued as needed by the System Administrator. Maintenance of core system software for the individual end–user sites will be provided by the ADS node where the end user is registered. This will include all software upgrades and additions.

Maintenance of their data holdings and documentation is the responsibility of each ADS node.

15–5 System Usage

A user of the system will enter ADS via a copy of the User Agent, either on his personal workstation, on his institutional computer, or remotely at a node. The system will provide for network and dial–in access from the user's home institution.

Through the User Agent (and transparently to the novice user) the investigator will have access to the databases, data descriptions, and documentation in the Directory Services. These services will be augmented in the near future with visualization, data reduction, and other analysis tools. The documentation in the system will include descriptive information on system services, available catalogs and other data sets, facilities, processing capabilities, data formats, and associated processing tools.

The information returned by the documentation services allows the scientist to determine the existence and location of the data and analysis software needed for his research. The user agent provides the user with the capability to query the system via Structured Query Language (SQL) or Query by Example (QBE). The user does not require an understanding of either of these languages; the system will prompt via menus or examples to obtain all needed information for the query. Once the desired data sets are located for interrogation, the system will provide on-line browsing capabilities to refine selections and to locate alternative data sets in other spectral bands.

Catalogs available on the system for the initial release include: The IUE and SAO Observation logs, the SAO Star Catalog, the IRAS Point Source catalogs and IRAS Additional Observations log, the Fourth UHURU Catalog, the Yale Bright Star Catalog, the Fourth Cambridge Radio Survey, the Burbidge *et al.* QSO Catalog, the HST Guide Star Catalog (GSC), *etc.* The initial list of data sets which will be included in the ADS contains over one hundred catalogs. However, it should be remembered that the purpose of the ADS is not to put catalogs and other data on-line but rather to make existing on-line data accessible. It is the charter of the mission data centers and organizations such as NSSDC to provide data, not of the ADS.

15–6 ADS Service Upgrades and Additions

As a matter of policy, new software tools development for inclusion in the ADS will be selected from proposals to the Astrophysics Software and Research Aids Program of NASA OSSA or developed by the ADS working group if they are core services or system enhancements.

15–7 Further Information

NASA has set up a project office at the Infrared Processing and Analysis Center at the California Institute of Technology. For further information on this program contact one of the following:

Dr. Guenter Riegler
Branch Chief, Science Operations
Astrophysics Division, Code SZ

NASA Headquarters
Washington, DC 20546

Dr. Frank Giovane
ADS Program Manager
Astrophysics Division, Code SZ
NASA Headquarters
Washington, DC 20546

Dr. John C. Good
ADS Project Manager
Infrared Processing and Analysis Center
MS 100-22
California Institute of Technology
Pasadena, CA 91125

Acknowledgements

This work would have been impossible without the tireless efforts of the ADS Working Group. Although there have been valuable contributions from a number of people in the extended group, most of the credit goes to the core subset listed below:

Dr. Edward Brugel (CASA, U. of Colorado, Boulder CO)
Alan Farris (STScI, Baltimore MD)
Dr. John Good (IPAC, CalTech, Pasadena CA)
Dr. Barry Jacobs (NSSDC/GSFC, Greenbelt MD)
Dr. Steven Murray (SAO/Harvard, Cambridge MA)
Dr. John Nousek (Pennsylvania St. Univ., University Park PA)
Dr. Peter Ossorio (U. of Colorado, Boulder CO)
Dr. Richard Pomphrey (IPAC, CalTech, Pasadena CA)
Lowell Schneider (Ellery Systems, Inc., Boulder CO)
Peter Shames (STScI, Baltimore MD)
Geoffrey Shaw (Ellery Systems, Inc., Boulder CO)
Randy Thompson (IUE/GSFC, Greenbelt MD)
Dr. Michael Van Steenberg (NSSDC/GSFC, Greenbelt MD)
Dr. James Weiss (JPL, Pasadena CA)

16

The NSSDC Services

Michael E. Van Steenberg & James L. Green
NASA Goddard Space Flight Center,
National Space Science Data Center,
Code 930.2,
Greenbelt, MD 20771, USA

16-1 Introduction

The National Space Science Data Center (NSSDC) has accumulated, over the last 20 years, several 100 Gigabytes of astrophysics and astronomy data sets and catalogs in digital form and thousands of feet of film. The astrophysics data in the NSSDC archive comes from such international spacecraft missions as the International Ultraviolet Explorer (IUE) and the Infrared Astronomical Satellite (IRAS). In addition to the large astrophysics data holdings, there are over 580 astronomy catalogs of stars and galaxies in the NSSDC archive. The NSSDC is a long term archive which has the responsibility of being a permanent NASA facility. The NSSDC data holdings are undergoing periodic media replacement. For instance, the IUE and IRAS data holding, currently stored on magnetic helical scan tape and 9 track magnetic tape respectively, are being rewritten to random access optical disk. In addition, the NSSDC is in the process of ensuring that all data in its archive has a backup which is stored at a separate location.

In the past several years the NSSDC has been migrating more and more of its astrophysics data and catalog archive to on-line or near-online status such that it is readily accessible electronically over national and international computer networks. The NSSDC supports electronic access from all the major computer networks such as the Internet (TCP/IP), SPAN (DECnet), BITNET (RSCS), and the international public packet switching network (X.25).

In the next 10 years, NASA will be launching many astrophysics spacecrafts whose data will be accessible through a variety of services at the NSSDC for NASA researchers and through the World Data Center-A for Rockets and Satellites (WDC-A R&S) for the international science community. The NSSDC is the home for the WDC-A R&S. There are now 36 world data centers internationally in the United States, Europe, the Soviet Union, Japan, and China.

A comprehensive guide to the services of the NSSDC is now available upon request (Green, 1990a). In addition, the document by Kim (1988) describes the NSSDC astrophysics archive data holdings, and the one by Warren et al. (1990) provides a brief description of the astronomy catalog archive. In this article we will give a brief overview of the NSSDC astrophysics services which provide scientists from all over the world access to both online and offline astrophysics and astronomy data and information in the vast NSSDC archive.

16–2 Current services

2.1 Astronomical Data Center

In this computer age, having astronomical catalogues in computer accessible form is very important to many astronomical research projects. As a result the Astronomical Data Center (ADC) at the NSSDC, one of six international astronomy data centers, works to provide the astronomical community with machine readable astronomical catalogs. These catalogs range from the well known such as the *Bonner Durchmusterung (BD)* and *Smithsonian Astrophysical Observatory (SAO) Star Catalog*, to the less widely known but important, e.g. the *Fifth Fundamental Catalogue (FK5)*. To service the astronomical community the ADC acquires, verifies, modifies, documents, and distributes astronomical catalogs and documentation in computer-readable form. Currently the ADC archives hold more than 580 computer-readable catalogs of astrometry, photometry, spectroscopy, and other miscellaneous data for both stellar and non-stellar objects. These data are acquired through exchanges with the Centre de Données astronomiques de Strasbourg (CDS), other astronomical data centers throughout the world, and by direct contributions from the members of the international astronomical community.

In 1989 the ADC has begun publishing astronomical catalogs on Compact Disk – Read Only Memory (CD–ROM). This extremely flexible and cost-effective distribution media can hold up to 640 Megabytes of data on a single disk. The first ADC CD–ROM has been published as a limited edition disk containing 31 astronomical catalogs in both plain ASCII text-file format and as standard FITS tables (Brotzman & Mead, 1989). The ADC is currently producing the first of a series of general distribution CD–ROM disks containing astronomical catalogs. This first CD–ROM pair (ASCII text-file and FITS) will contain approximately 100 of our most requested astronomical catalogs and is expected to be ready for distribution in the summer of 1991. We expect to produce more CD–ROM astronomical catalog disk pairs in the following years.

To assist the astronomical community in identifying what astronomical catalogs are available from the ADC, a menu-driven program is available on the NSSDC VAX computers that allows users to browse through descriptions of all the ADC catalogs and interactively place requests for desired catalogs. In addition, for the past several years, the ADC has been using computer networks to distribute astronomical catalog data and documentation held in its archives, thus avoiding the hassles and time delays involved in sending magnetic tape through the mail. This service is available through the NASA Science Internet (NSI) supporting the major national and international computer networks as well as dial-up modem lines.

2.2 Master Directory

Another operational system that is of general interest to scientists is the Master Directory or MD. This online search system provides brief overview information about NASA and important non-NASA space and Earth science data, archives, and other information. In many cases the directory offers automatic network connections to online information systems at locations throughout the world. The systems that MD connects to provide much more detailed information and access to the data of interest to the requestor. Although MD currently has far more information about Earth science data and archives it does support the astrophysics discipline. It is anticipated that a more comprehensive astrophysics information system will become available once the Astrophysics Data System (ADS) becomes operational (see chapter 15, this book). At this point MD will provide pointer and automatic links to the ADS capability as they become available.

The user of MD may search for data of interest through a variety of methods such as measured parameter, science discipline, location or spatial coverage, overall time period, spacecraft (or data source), sensor, investigator, campaign or project. For example, in the Astronomy science discipline area, MD also supports the sub-disciplines of X-ray, cosmic ray, gamma ray, ultraviolet, visible, infrared, and radio. The information displayed by the directory includes a descriptive title, summary abstract, key references, persons to contact, archive information, storage media information, and the values associated with the search keywords mentioned above. If a connection to another system with more detailed information is available, the connection can be invoked through the use of a simple LINK command. The use of the MD system over the last year has been exceptional and growing steadily with several 100 users per month using the system.

2.3 Data Acquisition, Archiving and Distribution

The NSSDC has a staff of scientists who are responsible for interfacing with the science user community and with the various NASA spaceflight missions. Each NASA mission, including astrophysics, must complete a document called the Project Data Management Plan (PDMP). This document is signed off by the NASA project officials, Headquarters, and the NSSDC and is designed to describe the mission's data products, how they will be archived and managed, and what facilities or data centers will be responsible for the archived data. It is through the PDMP that the NSSDC has a window on the future as to what missions and how much data the NSSDC will be archiving. Over the next several years the NSSDC will be receiving data from ROSAT, GRO, COBE, and many more important astrophysics missions.

Reviews of the PDMPs are performed regularly by the data suppliers, archiver, and users. Each of these three groups play important roles in the long term management of any useful archive. The archiver must ensure that archived data is safe, usable, and distributable. The scientific community must provide independent evaluation and guidance as to the scientific importance and usefulness of the archived data. Finally the data originators must provide the data and the scientific expertise. Regular reviews of these archiving plans help ensure the long term usefulness of the astronomical archives held at the NSSDC.

Advanced planing for data management and handling is essential for any archiving efforts. A scientist desire to put data in an archive is only the first of many steps in the data archiving process. Immediately the two questions that confront the archivist: what data ought to be archived; and what data needs to be archived. The former is a reflection of our obligation to future generations of scientist while the latter derives from the scientific consideration of what is needed (or required) to properly interpret the archived data.

Once these two basic questions are answered the practical day-to-day details of the archiving process need to be worked out, *i.e.* data format, data rights, storage media, data distribution, archival data rate, indexes, dataset total size, *etc.* These and other similar details must be established before any of the data can be properly archived.

Equally important is the documentation describing the scientific data. In our role as archivist we try to help and encourage the data suppliers to produce useful documentation about their data. Traditionally, documentation about archived data has been produced as hardcopy manuals. Thus as the number of datasets increase, the hardcopy documentation becomes more and more difficult to maintain and to distribute. In order to make these problems manageable we are in the process of converting our hardcopy documentation to computer readable form so that we can distribute as much of the documentation as possible via the same media that we distribute the science data, thus helping to keep data and associated documentation together.

The NSSDC supports the distribution and archiving of astrophysics data through a variety of online and offline services. There are currently over 18 online interactive systems that are supported (see Green, 1990b). The NSSDC is working with the Astrophysics Data System (ADS) program to ensure that the accessibility of its archive holdings will be available through this new data system.

2.4 Standards Office

The NSSDC supports the NASA Science Data Systems Standard Office (NSDSSO). This office was established to promote and facilitate the Space and Earth science communities to evolve cost-effective, interoperable data systems. The office has four distinct functional areas: standards administration, library, accreditation, and conformance and support. Briefly, the standards administration effort provides an active interface to other standards organizations within and outside of NASA to foster both the exchange of standards information and the development of new standards. The library is concerned with collecting, updating and disseminating information about existing and emerging standards of relevance to NASA data systems. The accreditation operation is concerned with the establishment, maintenance, and use of standards. Finally, the standards conformance and support operation is concerned with providing a support to users trying to use a recognized standard.

An important rôle of the NSDSSO is that it supports the Flexible Image Transport System (FITS) format (see related chapter on page 253, this book). The office supports an online interactive system which contains abstracts of the appropriate documentation of the FITS standard format. Within NSDSSO a FITS standard office, staffed during the work day, has been established which assists requesters in the use of FITS in their products, participates in the evolution of FITS, and has been developing FITS conformance software in consultation with FITS experts. In addition, the FITS Office is currently documenting the current understanding of FITS, thus presenting FITS as a well illustrated and supported standard.

16–3 Future services

3.1 Astrophysics Data System

The NSSDC is one of the six initial nodes in the Astrophysics Data System (ADS) (see chapter 15, this book). The NSSDC contributes to the ADS in several ways. Possibly, the most important to the scientist is the long term archiving and distribution of astrophysics data. Other contributions include: providing on-line access to the ADC and MD as well as to the astronomical catalogs held by the ADC; providing the underlying network distributed database system, the Distributed Access View Integrated Database (DAVID) system software; finally the NSSDC node acts as the access point for non-astrophysics users to the ADS.

Results of ongoing research software development projects at the NSSDC in the following fields will be supplied to the ADS as they are available: data browsing tools, Artificial Intelligence and expert systems, database management systems, data compression, and data visualization.

3.2 Data Archiving and Distribution System

In response to the tremendous success of the 'fast' distribution of International Ultraviolet Explorer data (spectra extracted usually within 24 hours and highly processed via USSP in a few minutes, see chapter on IUE, this book), the distribution of data via the international networks, and the steadily increase in demand for astronomical catalogs, we are creating a system into which nearly all of the publicly available astronomical data held at the NSSDC will be held on-line or near-online (less than 5 minutes to transfer to on-line status). In order to accomplish this we are using the basic building blocks of the Space Telescope archiving system and generalizing them to handle a much wider range of data products. This includes future data products such as from the Gamma-Ray Observer (GRO) and the Far Ultraviolet Spectroscopic Explorer (Lyman FUSE) as well as the past and present data archives from the Infrared Astronomical Satellite (IRAS), International Ultraviolet Explorer (IUE), Orbiting Astronomical Satellite 3 (OAO-3 Copernicus) spacecraft. In addition, all of the astronomical catalogs in the ADC archives will be available via the same system. Even though we are responsible for the ultimate archiving of NASA's astronomical archives we also act as a secondary archive for non-NASA data that is needed or is of interest to the astronomical community using NASA's archived data.

All of the actively requested astronomical data held at NSSDC will be available via the NSSDC Data Archiving and Distribution System (NDADS). Depending on the volume of data and the media requested (*i.e.* 9-track or 8mm tape, network) we expect to satisfy all reasonable requests within 24 hours. Of course more modest requests for distribution via the networks should be satisfied much more rapidly (less than 5 minutes to start network transfer).

For the past few years the only media that was generally suitable for data distribution was 9-track magnetic tapes. With the completion of the NDADS facilities we will also be able to support routine distribution on other media: network (DECnet and TCP/IP), 8mm tape, Digital Audio Tape, re-write optical, and Write Once Read Many optical disks.

By using WORM optical disks to store the astronomical archive we have reduced several of the burdens associated with maintaining a useful data archive. Optical disks provide much more storage space and require much less maintenance to ensure the archive data integrity (*e.g.* no end-to-end spinning; expected life times larger than 100 years; reduced constraints on environmental conditions).

During our initial setup and testing of the NDADS facility, the IUE and IRAS data archives will be loaded onto optical disks and placed into a robotic optical disk 'jukebox'. This will allow us to remove the *middle man* (the person that physically picks up the tape from a tape rack and mounts it on a tape drive) involved in most tape based data archives. Since these are our most popular large astronomical archives they will allow us to provide a fast access to significantly large data archives and provide us with valuable experience in this new system. As a result, NDADS will allow us to place the full astrophysics archive on a medium that does not require active maintenance.

In addition the NDADS facility at NSSDC will be the basis for the archiving database management and distribution services for the currently forming High Energy Astrophysics Science Archive Research Center (HEASARC), while the scientific and user support will be through the Laboratory for High Energy Astrophysics at Goddard Space Flight Center.

16–4 How to access the NSSDC archives

The NSSDC supports the distribution and archiving of astrophysics data through a variety of online and offline services. There are currently over 18 online interactive systems that are supported (see Green, 1990b). Current and future online services of interest to the astronomical community are summarized in Tables 16–1 and 16–2.

Current Online Data System	Node[a]		Purpose
ADC	nssdca	nodis	Access to the ADC online catalog request system.
EXOSAT	ndads	xray	Access to the EXOSAT Database system.
IUE	nssdca	nodis	Electronic requests for IUE archived data.
IUE ULDA	nssdca	nodis	Access to the US national node for the IUE Uniform Low Dispersion Archive.
MD	nssdca	nodis	Access to the NASA Master Directory system.
ROSAT MIPS	rosat	[b]	Access to the US ROSAT Mission Information and Planning System.
SIS	nssdca	nodis	Access to the Standards Information system.

Table 16–1: Current online services of interest to the astronomical community

[a]Node names may be used in either of two ways: as a SPAN node name, and as an Internet name with the addition of ".gsfc.nasa.gov" (i.e. nssdca is both nssdca:: and nssdca.gsfc.nasa.gov).

[b]Please contact Jeanne Behnke at: Behnke@rosat.gsfc.nasa.gov or ROSAT::BEHNKE for the username and required password.

For current information, interested individuals are invited to subscribe to the NSSDC quarterly newsletter, by sending a request to the NSSDC request service given below.

Researchers can obtain more information about the NSSDC's data archive, how to contribute to it or how to request data from it (including cost and availability concerns), as well as data services by addressing their questions as follows.

Future

Online Data System	Node	Purpose
ADS	ndads	Access to the Astrophysics Data System.
HEA-SARC	ndads	Access to the High Energy Astrophysics - Science Archive Research Center.
NDADS	ndads	Access to the NSSDC Data Archive Distribution System.

Table 16–2: Future online services of interest to the astronomical community

For information or data requests please contact:
    From within the USA:
    Request
    National Space Science Data Center
    Code 933.4

    From outside the USA:
    Request
    World Data Center A for Rockets and Satellites
    Code 930.2

    NASA Goddard Space Flight Center
    Greenbelt, MD 20771
    Telephone: (301) 286-6695
    FAX: (301) 286-4952
    Telex: 89675 NASCOM GBLT
    TWX: 7108289716
    SPAN: NCF::Request
    Internet: Request@nssdca.gsfc.nasa.gov

For data submissions from within the United States please contact:
    Dr. H. Kent Hills
    National Space Science Data Center
    Code 933.8
    NASA Goddard Space Flight Center
    Greenbelt, MD 20771
    Telephone: (301) 286-4106
    FAX: (301) 286-4952
    SPAN: NCF::Hills
    Internet: Hills@nssdca.gsfc.nasa.gov

and for submissions from outside the United States please contact:

Dr. James L. Green
World Data Center A for Rockets and Satellites
Code 930.2
NASA Goddard Space Flight Center
Greenbelt, MD 20771
Telephone: (301) 286-7354
FAX: (301) 286-4952
Telex: 89675 NASCOM GBLT
TWX: 7108289716
SPAN: NCF::Green
Internet: Green@nssdca.gsfc.nasa.gov

References

[1] Brotzman L. E., Editor (1990), "Astronomical Data Center", *BAAS*, **22**, No. 2, 914.
[2] Brotzman L. E. & Mead J. M. (1989), "Astronomical Data Center", *BAAS*, **21**, No. 2, 811.
[3] Green J. L., Editor (1990a), "A guide to the National Space Science Data Center", NSSDC 90-07, June 1990.
[4] Green J. L. (1990b), "The new Space and Earth Science information systems at NASA's archive", Government Information Quarterly Symposium Issue on NASA, **7**, 141.
[5] Kim S. J. (1988), "Descriptions of astronomy, astrophysics, and solar physics spacecraft, investigations, and data sets", NSSDC/WDC-A R&S 88-12, Vol. 5A and 5B, June 1988.
[6] Warren W. H. Jr., Mead J. M., Brotzman L. E. (1990), "Status Report on Machine-Readable Astronomical Catalogs", Astronomical Data Center, September 6, 1990.

17

The Space Data Centre at RAL

David Giaretta & Eric Dunford
Rutherford Appleton Laboratory
Chilton, Didcot
Oxfordshire OX11 8QT, United Kingdom

17-1 Introduction

The Rutherford Appleton Laboratory (RAL) is funded by the United Kingdom Science and Engineering Research Council (SERC) to provide scientific support services to the UK academic community in several areas. Of interest in the context of this book are the services to the UK astronomical community.

Chapter 18 of this book discusses the Starlink astronomical database system, SCAR, which is being developed and maintained at RAL. This chapter discusses several other services that are provided by RAL, both in terms of the on-line data and nearly on-line data, and equally as important, the supporting expertise that goes along with this, both in the Space Data Centre itself and also in the Starlink system.

The Space Data Centre (SDC) was formed at the Rutherford Appleton Laboratory from a consortium of space related projects in the astronomy, solar terrestrial physics and earth observation branches of space science. Its purpose is to act as a focus for the space data acquisition and processing activities in which RAL is involved. A levy on each of the user projects provides most of the financial resources required to maintain it and support new developments in hardware and software. The demands of different projects are very different from one another depending on the field of study, international and national collaborators involved and the source of the data. Thus the Space Data Centre will be called on to carry out different roles ranging from postman to providing comprehensive data reduction and analysis facilities. One important function is however that it acts as a point of contact between RAL and academic communities and international agencies and projects, such as the Consultative Committee for Space Data Systems (CCSDS), and European Space Information System (ESIS) (see related chapter this book, page 127). To this end it has acted as a catalyst, together with Starlink, in setting up international data links with Europe and the USA. Thus international gateways are now provided at RAL into ESANET and the ESIS pilot system.

17-2 Facilities at the SDC

In the following we will describe the astronomically related activities at the SDC. RAL provides a processing centre for satellite data, which are then further distributed throughout the UK. Data are also received from international collaborators, for which the SDC is a secondary distribution centre. It is also anticipated that in many cases, after a period, much of the data will be re-processed and re-distributed to users. In the process a great deal of expertise is built-up about data products and their production and methods of analysis, involving tools developed both at RAL and elsewhere.

The SDC computer hardware is DEC VAX/VMS based, supported by mass storage media such as SONY 6.4 GByte WORM optical disks, Exabyte tape, as well as 0.5 inch CCT. Note that compatibility is essential between the SDC and Starlink, which may mean an eventual move to UNIX. Moreover the SDC must support the media on which data is received from other sources.

It is worth noting here the inter-dependence of hardware and software requirements in the following sense. The availability of new software packages can make different datasets of interest to the user; this usually means that users can start their processing nearer to the the raw data, which usually implies using a greater volume of data, and hence the need for greater volumes of on-line or nearly on-line data. On the other hand new storage devices, both random access and sequential, costing far less per megabyte of data than in the past, allow developers of analysis tools for the general community to be more imaginative in their use of the data.

17-3 Data available or expected to be available

There are many datasets connected with satellite projects with which RAL has had direct involvement. We discuss a number of these, specifying the original source of the data, the access methods and tools available. In the next section we also mention future developments in these and other datasets.

3.1 IRAS

The InfraRed Astronomical Satellite was a joint NASA, SERC and Dutch project and has been described in chapter 3. The main post-mission processing centre, the Infrared Processing and Analysis Center (IPAC), at Pasadena send full copies of the Calibrated Raw Detector Data (CRDD) to RAL, while data from the Low Resolution Spectrometer (LRS) are received from Holland.

CRDD data. At the SDC the aim is to make CRDD data covering several degrees of sky available within a few days, following receipt of a data request at the account RLSTAR::IRASMAIL specifying the area centre, extent and number of bands desired.

LRS data. LRS spectra are available in two forms. The first consists of the LRS Atlas, a carefully selected sub-set numbering 5425 of the full database of spectra, which may be used within the SCAR system, described elsewhere. The second form is the full LRS database, containing up to 170,000 spectra. This is most easily accessed using the IRASLRS account on RLSTAR containing software for selecting and combining the sometimes rather noisy individual traces of objects.

3.2 IUE data

Several forms of IUE data are available via the SDC and/or Starlink, ranging from raw data, fully extracted spectra and catalogues. The IUE Uniform Low Dispersion Archive (ULDA) which has been fully described in chapter 5 is available on-line. A selection of the raw, photometrically corrected and extracted data files may be requested in an interactive system (IUEDEARCH on RLSTAR) for batch processing overnight. A catalogue inquiry system specific to IUE allows the user to select appropriate images. It is expected that, in the next few years, after the IUE data has been re-processed, much more IUE data will be on-line via an optical disk jukebox.

3.3 SMM data

Solar Maximum Mission (SMM) data are available off-line, copied by special request onto disk, to be transferred to a user's home computer.

3.4 ROSAT Wide Field Camera (WFC)

The ROSAT data system has been fully described in chapter 1. The initial processing of the Wide Field Camera data is performed at the SDC and subsequently distributed to various groups for detailed analysis. An archive of data is kept on optical disks.

3.5 STADAT catalogues and spectral atlases

The data holdings available on-line on the Starlink database (STADAT) microVAX are managed by SDC staff. They consist of a variety of astronomical catalogues, mostly supplied by the Centre de Données astronomiques de Strasbourg (CDS) and formatted to allow access using the SCAR system. There are also several spectral atlases. More data are being put on-line as requested by users and as disk space allows. It is expected that very soon a great deal more data will be made available on-line, using optical disk devices.

3.6 Data Organisation

Data sets are stored in a great variety of ways, for example: tapes, both IBM standard label special format, as well as unlabelled FITS tapes; disk, CD–ROM, erasable optical platters, WORM platters in VMS Files-11 format and associated paper documentation.

Detailed descriptions of the archive data available may be found in the Starlink documentation, available on all Starlink nodes, or requested from RLSTAR::STAR.

17–4 User services

Starlink software is widely used in the U.K. for analysis of astronomical data, particularly the ADAM environment, however other analysis environments are available, such as MIDAS, IDL and IRAF, with limited support. Database packages such as SCAR, mentioned above, and R-EXEC are used to access catalogue data, although commercial systems are currently under evaluation.

The typical user of the RAL data systems gains access through SPAN or through the Joint Academic NETwork (JANET). A limited number of international users may be allowed with prior authorisation.

17–5 Future plans

New data acquisitions are expected over the next few years. For example all IUE data are to be re-processed, starting in early 1991, and a copy of the new products will be available in the SDC, possibly on-line. Data from certain instruments on ISO, ROSAT and SPECTRUM-X may be archived in the SDC, as will new IRAS data products produced by IPAC, and data from ERS-1, SOLAR-A, SOHO, CLUSTER, CRRES, FREYJA and Polar Platform will also pass through the SDC for processing at one level or another.

18

Database applications in Starlink

Clive Davenhall
Royal Observatory
Blackford Hill
Edinburgh EH9 3HJ
United Kingdom

18-1 Introduction

Starlink is a computing facility provided by the United Kingdom Science and Engineering Research Council (SERC) for the use of UK astronomers. It was established in 1980 to provide facilities for the efficient reduction of the increasing volume of digital data generated by modern telescopes and detectors, and its primary purpose remains the interactive reduction and analysis of observational data. It is managed by the Rutherford Appleton Laboratory (RAL). The facilities provided divide naturally into hardware and software.

The hardware consists of a network of computers with nodes at most of the astronomical institutions in the UK. Traditionally the type of hardware used has been the DEC VAX running the VMS operating system. STADAT, a microVAX II located at RAL, is a node of the network particularly associated with database activities. This machine has a large amount of disk space, used to hold a large collection of catalogues and databases, which can therefore be accessed on-line.

The software consists of the *Starlink Software Collection*, a collection of software for the reduction and analysis of astronomical data. Starlink maintains and distributes the collection and it is developed both by effort from within the Starlink facility itself and from effort available in the astronomical community. The database software in the collection is available throughout Starlink, though other nodes will have fewer catalogues available than STADAT.

This chapter will concentrate on general database activities in Starlink. It will specifically exclude the various observatory and instrument archives that are available, such as those for the IUE satellite or the La Palma Observatory, though often these may be accessed through Starlink. Some of these archives are discussed in other chapters (see, for example, chapters 5 and 12). Once the observation archives have been excluded, the databases that remain are mostly catalogues of various sorts. These catalogues divide into broadly two types:

- The computer–readable versions of published star catalogues, such as those distributed by the International Network of Astronomical Data Centres. These catalogues are usually compilation catalogues (CCs) of various sorts in the terminology of Jaschek (1989).
- Private catalogues created by an individual researcher or group for some specific purpose. These catalogues may vary from, for example, a list of positions and magnitudes for a few objects to a catalogue of the positions, magnitudes and other parameters for all the images detected in an automatic scan of a wide-angle Schmidt plate, perhaps amounting to over a million stars and galaxies. Though the details vary, these catalogues are generally closer to observational catalogues (OCs) in Jaschek's terminology.

Catalogues of both of these types are usually tabular in form, listing a set of measurements for the same set of parameters (such as positions, magnitudes, *etc.*) for a list of objects. For example, table 18–1 shows an extract from the Fourth edition of the Yale Bright Star Catalogue (Hoffleit & Jaschek, 1982), listing a few of the parameters for a few stars.

HR	Name	RA (1900)			Dec (1900)			V	(B–V)	...
		h	m	s	°	′	″	m	m	
⋮	⋮	⋮			⋮			⋮	⋮	
101	10 CET	0	21	29.6	-0	36	12	6.19	0.90	...
102		0	22	14.0	-26	6	1	5.98	1.03	...
103	47 PSC	0	22	50.0	17	20	21	5.06	1.65	...
104		0	22	51.1	43	50	29	5.17	0.03	...
⋮	⋮	⋮			⋮			⋮	⋮	

Table 18–1: An example of a tabular dataset; an extract from the Yale Bright Star Catalogue.

It is worth discussing the catalogues derived from automatic scans of Schmidt plates in some detail because they tend to be the largest catalogues routinely processed and hence make the most extreme demands on the database software. Photographic plates obtained with wide-angle Schmidt telescopes such as the UK Schmidt in Australia, the ESO Schmidt in Chile or the Palomar Schmidt contain a prodigious amount of information. Plates from the UK and Palomar Schmidts are 35.6 cm^2 and may contain images of hundreds of thousands or even millions of objects. Each of these telescopes produce some hundreds of plates during a year of normal operation. The difficulty of analysing all the information present on these plates, the so-called 'Schmidt Problem', has long been recognised (see *e.g.* Fellgett, 1979). The only effective way to extract all the information from the plates is to scan them automatically using fast microdensitometers such as the COSMOS machine (MacGillivray & Stobie, 1984) at the Royal Observatory Edinburgh (ROE). The digital information produced by these machines is then automatically analysed to detect and produce parameters (usually including positions, magnitudes, orientations, ellipticities, *etc.*) for all the star and galaxy images detected on the plate. Typically, the catalogue produced by such a measurement could contain half a million objects and the fast microdensitometers can complete a measurement in a few hours.

Thus large numbers of large catalogues[1] are generated and the database software must be capable of manipulating them efficiently, often including joining them to produce even larger 'survey' catalogues covering a wide area of sky. In practice many of the catalogues are processed by specialised software, but there remains a significant requirement to process them with a flexible database package. Fast microdensitometer scans of Schmidt plates are not the only source of large catalogues: for example, some of the catalogues produced by analysis of the results of the IRAS satellite are of a similar size. Furthermore, large catalogues are likely to become increasingly common: new and larger Schmidt telescopes are being considered or planned, more and faster measuring machines, such as the SUPERCOSMOS machine at the ROE, are under development and other branches of astronomy are increasingly producing large catalogues. The requirement to efficiently process large catalogues dominates the design of the database software.

The principal database package available in the Starlink software collection is the *Starlink Catalogue Access and Reporting (SCAR)* system. In addition to SCAR a number of other packages are also available, but these are more specialised, less flexible or in some way more limited. SCAR will be discussed in the following section and the other packages in the subsequent one.

---

[1] The software for the COSMOS machine determines thirty-two parameters for each object detected, each stored as a four-byte REAL number. Thus a catalogue of half a million objects will occupy sixty-four Mbytes.

18-2 SCAR

2.1 History and nomenclature

The Starlink Catalogue Access and Reporting system (SCAR) is a relational database management system for astronomical catalogues. SCAR comprises a number of applications for performing database operations on the catalogues, as well as defining a format for the catalogues and providing a set of utilities to access catalogues in this format. The defined format of the catalogues and the routines to access them are together referred to as the *Flexible Astronomical Catalogue Transport System* (FACTS). FACTS was formerly called *Astronomical Data Catalogues* (ADC), but the name was changed to avoid a clash with the acronym for the Astronomical Data Center in the United States, though the acronym ADC persists in internal Starlink documentation. The original, rather limited, design and implementation of ADC was developed around 1980 by D.J. Carnochan and S.L. Wright of University College London, and K.P. Tritton and D.R.K. Brownrigg of the ROE, and was primarily intended for use with the compilation catalogues (CCs) distributed by the International Network of Astronomical Data Centres. The SCAR package largely dates from the work of J.H. Fairclough of the IRAS Post Mission Analysis Facility (IPMAF) at RAL (see also chapter 3), who extended the ADC design, re-implemented it and wrote the basic set of SCAR applications. This first version of SCAR was orientated towards catalogues derived from the IRAS satellite. It was released in 1985 and ran under the Interim Starlink Environment. Subsequent work has involved converting SCAR to use the more sophisticated ADAM environment (see Lawden, 1989, for a description of ADAM) and adding more applications. IPMAF continue to be responsible for the maintenance and development of SCAR. A set of catalogues is available for use with SCAR. These include versions of catalogues distributed by the International Network of Astronomical Data Centres converted to FACTS format and catalogues of results from IRAS.

2.2 Requirements

A database package designed for general use in astronomy must satisfy a number of requirements. It must be capable of handling datasets that range in size from small, private lists to the large catalogues derived from Schmidt plates discussed in the previous section. The operations that are carried out on the various sorts of catalogues are remarkably similar and include the following broad classes of operations.

Calculations Deriving new quantities from the existing parameters for all the objects in a catalogue. An example might be calculating stellar

Database applications in Starlink

temperatures from a set of magnitudes and colours.

Selections Selecting all the objects in a catalogue which satisfy some set of criteria (or, indeed, the alternative set which do not satisfy the criteria). A simple example might be selecting all the stars in a catalogue brighter than a limiting magnitude.

Reporting Displaying the information in the catalogue, either in the form of listings or graphically.

Joining Combining two or more catalogues by finding the objects that they have in common. Often, but not always, objects in two catalogues will be paired on the basis of similar positions.

Sorting Arranging the objects in a catalogue so that the values for some parameter occur in ascending or descending order. For example, a catalogue might be sorted so that objects are ordered in increasing Right Ascension.

Flexibility is most important. The format and contents of the catalogues vary, as do the operations that are to be performed on them. Particularly for new datasets, the user needs to be able to interactively perform operations, display the results and in the light of the results perform fresh operations. Sequences of operations (such as calculate new parameters, perform selections on a mixture of the old and new parameters, display the results and then perform fresh calculations) are common.

Speed and efficiency in accessing the data are important, particularly for large catalogues. In practice this requirement means that the organisation of the catalogue must be kept simple. Standard commercial database applications are generally concerned with accessing large but fairly static datasets. By comparison astronomical applications frequently handle volatile datasets where values must frequently be written and updated. Thus efficiency in writing new values is as important as efficiency in retrieving existing ones.

2.3 Database design

SCAR is a relational database. Relational databases represent their data as a two-dimensional table or *relation* with the following properties (see any standard text on databases, for example Kroenke, 1977).

1. Each entry in the table is single-valued: repeating groups, arrays or complex structured items are not allowed.
2. The entries in any column of the table are all of the same kind.
3. Each column has a unique name and the order of the columns is irrelevant.
4. No two rows in the table are identical.
5. The order of the rows is insignificant.

The columns in the relation are referred to as *attributes* and the rows as *tuples*. Relations of this sort are a perfectly natural description of astronomical catalogues; the parameters of the catalogue are the columns or attributes of the relation and the objects it contains the tuples. For example in Table 18–1 the HR number, Right Ascension, Declination, V magnitude, (B–V) colour, *etc.*, are the attributes and each star is a tuple. In SCAR each tuple is implemented as a separate record in a disk or tape file for efficiency of access and each attribute is a field in the record. In SCAR, and for the remainder of this chapter, the more familiar nomenclature of field and record will be used rather than attribute and tuple, though it should be remembered that a physical record is merely the way that SCAR chooses to implement a tuple. In astronomical catalogues each record usually corresponds to a different object and 'object' will be used synonymously with record for the rest of the chapter[2].

The key to the flexibility of SCAR, and indeed of relational databases in general, is referring to fields by name. Generalised database applications can be written in which the user supplies the names of the fields and specified operations are then performed on these fields. For example, if a catalogue contained fields of B and V magnitudes and the user wished to compute the colour (B–V), he would run a general application for performing calculations, specify the fields concerned and the operation (subtraction in this case) and the colour would be computed. Thus, the necessity for writing a special program for every calculation is avoided.

In addition to storing the data for the catalogue, the records and fields that comprise it, each catalogue must also store the names of the fields and sufficient information about their position and format in the record to allow values to be retrieved from them. In SCAR the data for the catalogue and the description of the catalogue are held in separate files, usually referred to as the data file and the description file respectively. In addition to the names and details of the fields, the description file also contains catalogue *parameters*; items that pertain to the entire catalogue, such as the number of records it contains, the record size, the title of the catalogue, the medium (tape or disk), *etc.* Some parameters, such as the record size and number of records, are mandatory, others are optional. Typically, the information for each field includes its name, position in the record, formats, (both internally in the catalogue and externally for presentation to the user), units, comments, *etc.*

SCAR allows considerable flexibility in the format of the data catalogue. It may be either a formatted text file suitable for printing, or a binary file for

---

[2]Throughout this chapter the word *object* is used in its astronomical sense of a heavenly body rather than in any of its computer science senses.

efficient access; it may have sequential access suitable for reading the entire catalogue or direct access suitable for retrieving individual records. Both tape and disk catalogues are supported. The options chosen for a particular catalogue depend on the use that is to be made of it. An individual field must have a type chosen from one of the standard types of Fortran 77 or one of a number of non-standard integer types. The type is specified using Fortran 77 style format specifiers.

SCAR has the ability to handle *null values*. These are special values, outside the range of genuine values for a field, which indicate that, for a particular object, no value is available for the field. For example, some objects in a list of photometry may not have measurements in all the colours, and the null value would be used for these missing measurements. The null value is not a constant throughout SCAR, but its value is defined separately for every field in a catalogue and, like the other properties of fields, it is specified in the description file.

The final important feature of SCAR is its ability to handle *index* catalogues. The catalogues described so far, consisting of tables where each row is a set of fields containing values for some object, are properly called *master* catalogues. An index catalogue consists of a description file and a data file, like a master catalogue. The data file may contain fields in the normal way, but at least one of the fields is a pointer, or index, identifying individual records in another catalogue. The catalogue to which these indices refer may be either a master catalogue or another index catalogue. Performing operations on catalogues will often generate an index catalogue rather than creating a new master catalogue. Specifically, index catalogues can arise in three types of operations:

Selection Here the indices point to those records in a master catalogue which satisfy some selection criteria.

Calculation If a quantity has been calculated for which there is no field in the original catalogue, the index will contain the values calculated for the new field and pointers back to the corresponding original record in the master catalogue.

Sorting If it is required to sort the catalogue into other than its original order, the indices are arranged in the required order but point back to the original records in the master catalogue.

As an example, Figure 18–1 shows the first few records of the index catalogue generated by selecting all the stars in the Bright Star Catalogue with a V magnitude brighter than 6.0. Note how the indices in the index catalogue correspond to record numbers in the master catalogue.

SCAR makes access to index catalogues 'transparent', so that the index catalogue and its master appear to the user identical to the equivalent

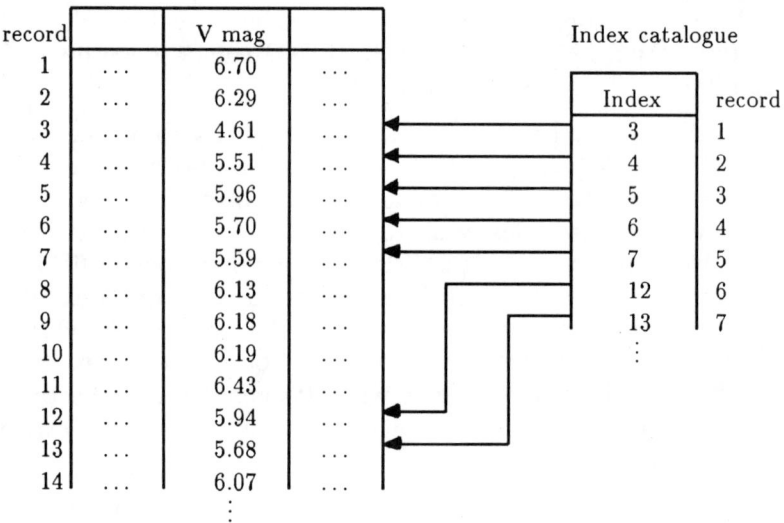

Figure 18–1: An example of an index catalogue.

master catalogue. Index catalogues are a compromise between the size of the catalogues and the speed with which they can be accessed. A set of index catalogues and a single master catalogue to which they point are much smaller than the equivalent set of master catalogues. Conversely index catalogues are, of necessity, slower to access because more operations are involved. Utilities exist to convert an index catalogue and its master into the corresponding master catalogue.

2.4 Facilities available

SCAR contains facilities to support operations in the five broad classes outlined in section 2.2:

Calculations New values for fields may be defined in terms of an algebraic expression involving existing fields. Following the previous example if a field COLOUR was to be computed from B and V magnitudes held in fields BMAG and VMAG respectively, the expression would be

$$COLOUR = BMAG - VMAG$$

Complex algebraic expressions involving many fields, constants, brackets and a variety of mathematical (and astronomical) functions are permitted. The new values computed may overwrite an existing field or create a new field in an index catalogue.

Selections Objects may be selected according to different sorts of criteria.

- Simple selections of all the objects with values larger (or smaller) than a given value in some field, for example, all the objects with a V magnitude brighter than 6.0. Several such criteria may be combined to form an arbitrarily complex selection in a single pass through the catalogue; for example, all the objects with a V magnitude in the range 5.0 to 6.0, which have a (B−V) colour greater than 0.1 and projected rotational velocities greater than 50 km/s, *etc*.

- All the objects that fall inside an arbitrarily shaped polygon defined in a plane formed by choosing any two fields as the axes. This feature is useful for separating objects which occupy different regions when plotted on a graph, for example, selecting main sequence stars in a H–R diagram.

- Selecting an arbitrary number of objects with the largest or smallest values for some specified field, for example, the hundred brightest objects in V. For efficiency this option is available without sorting the entire catalogue.

- Selecting every n^{th} object from a catalogue, for example selecting every hundredth object. This technique is useful for extracting an unbiased sample from a catalogue which is small enough to be experimented on interactively. The procedures and criteria developed interactively may then be applied to the entire catalogue.

Reports and plots   Information may be extracted from the catalogues in a variety of formats, either orientated towards displaying all the fields for an object or objects, or towards displaying selected fields for all the objects. This information may either be displayed interactively on a terminal, printed or saved as a disk file, perhaps to be processed by private programs. Facilities also exist to plot scattergrams and histograms, and celestial coordinates may be plotted using an Aitoff (all sky) or tangent plane projection.

Joining   Catalogues may be joined by identifying objects in common using either an exact match or an approximate match. In an exact match, objects with identical values for some field, normally a name, are identified as being the same object. In an approximate match, objects are matched by having similar coordinates, allowing some tolerance for error, that is the coordinates do not have to be identical. This process is computationally intensive and both catalogues must be sorted into the same order before it can be attempted.

Sorting   Catalogues may be sorted into ascending or descending order on any field. Also, they may be sorted so that objects with identical values for the primary sort field are further sorted on another field.

SCAR is able to handle fields containing positions measured in angular coordinates expressed in hours or degrees, minutes and seconds, and represented in a variety of formats. It also has functions to convert between equatorial, ecliptic, Galactic and supergalactic coordinate systems and to calculate precession and great circle distances. These functions may all be included transparently in expressions for calculating new fields and performing selections.

18–3   Other database systems

In addition to SCAR a number of other database systems are available on Starlink, though none is as powerful or flexible.

Haggis   is a relational database system for handling tabular catalogues very similar its in basic principles to SCAR. It was developed at the ROE, primarily for analysing COSMOS data. Compared with SCAR its catalogues are more restricted in format, it does not have the facilities for handling angular coordinates or sorting and it is slower. Development was suspended in 1986 to concentrate work on SCAR.

R-EXEC is a relational database system for general scientific rather than specifically astronomical work. It was developed at RAL and is widely used in geophysics. Currently it does not contain any specifically astronomical functions.

CHART is a specialised package for producing finding charts rather than a general database system. It accesses a number of standard star catalogues, held in a fixed format.

Finally, the database systems do not provide facilities for complex statistical analyses, but the commercial statistics packages Genstat 5 and Clustan are available for this purpose. It is possible to extract tables from the databases and import them into the statistics packages.

18–4 Comparison with HDS

This section compares the FACTS relational database format used by SCAR with the other major data format used by Starlink, the *Hierarchical Data System (HDS)*. Due to space considerations, it is not possible to give here an extensive description of HDS, for further reference, the related Starlink documents ar listed in a section below. HDS plays a central rôle in Starlink software for the reduction and analysis of observational data, being used to store two-dimensional images, spectra, and other datasets, both raw and in various stages of reduction. Thus, the datasets that it handles are considerably more complex than the simple, flat tables of FACTS.

As its name implies, HDS is a hierarchical system, with items in the dataset occupying points in the hierarchy. An HDS dataset is probably most easily visualised as a tree structure. Each item in the hierarchy may be either simple or structured. A simple item contains a simple datum or an array of values. Each simple item has a type, which may be one of the standard types of Fortran 77, or various sorts of integer number. A structured item contains one or more simple items or other, subordinate structures. All the items in a structure have a name, by which they are manipulated. Access to items is always by name and the way the data are stored is hidden from the application programs which manipulate them. Figure 18–2 shows an example HDS structure for a simplified dataset obtained from a CCD detector. Here NAME, RA and DEC are simple items, probably of type CHARACTER. DATA_ARRAY is also a simple item, but it contains a two-dimensional array of values read from the detector. It will probably be of type REAL or INTEGER. Conversely, DESCRIPTION is a structured item, with components NAME, RA, DEC and DATE.

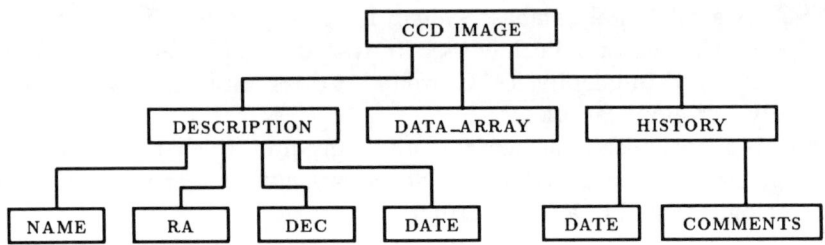

Figure 18–2: A schematic representation of a simple HDS structure.

The tree representing the HDS structure may be arbitrarily shaped, with an arbitrary mixture of structured and primitive items at any level. Items in the structure may be dynamically added and removed as the dataset it holds is processed. In order to allow related applications to cooperate in processing datasets a standard sort of HDS structure has been defined, the *N-dimensional Data Format* (NDF). This format specifies standard names and types for simple items and standard names and components for standard structured items, and where appropriate gives rules for processing them.

There are both similarities and differences between HDS and FACTS. Both achieve data abstraction by referring to data items by name. The FACTS system is appropriate for describing simple, flat, relational tables, whereas HDS can describe arbitrarily complex, hierarchical datasets. In FACTS the primitive data item is always a simple datum (a field in a particular record) whereas in HDS it may be a simple datum or an n-dimensional array. FACTS is restricted to simple, catalogue data, whereas HDS can represent more complex, structured datasets. It is, of course, possible to represent a flat, relational table using HDS and one of the future developments proposed for SCAR is to add the option of accessing catalogues stored as HDS rather than FACTS files.

18–5 Further reading

Internal Starlink documentation is available for the packages discussed in this chapter. Table 18–2 lists the relevant documents. Starlink documents are numbered in several series, amongst which the *Starlink User Notes* (SUNs) and *Starlink General Papers* (SGPs) are germane. In some cases these documents may refer to user guides which are not numbered in the Starlink series, but which are available separately. Further details and copies of the documents may be obtained from the Starlink Project, Rutherford Appleton Laboratory, Chilton, Didcot, Oxfordshire, OX11 0QX, UK.

Item	Document
SCAR	SUN70
Haggis	SUN69
R-EXEC	SUN97
CHART	SUN32
HDS	SUN92
NDF	SGP38
History of Starlink	SGP31

Table 18-2: Starlink documents describing the packages discussed in this chapter.

Acknowledgements

This chapter would not have been possible without the efforts of the people who contributed to the software it describes. SCAR and Haggis share a common ancestry and many of the people who have worked on them have contributed to both systems. The first released version of SCAR was largely the work of J.H. Fairclough. Other people who have contributed to these systems include S.M. Beard, D.R.K. Brownrigg, D.J. Carnochan, J.J. Cooke, J.P. Day, E.J. Gershuny, D.L. Giaretta, A. Hobson, B.D. Kelly, S.K. Leggett, C. Singh, B.C. Stewart, K.P. Tritton, H.J. Walker, A.R. Wood, S.L. Wright and the present author.

R-EXEC was developed, under the direction of B.J. Read, by the database section of the Central Computing Division at RAL. CHART was originally developed at the Royal Greenwich Observatory. The people who have contributed to it include P.M. Allan, L. Bell, J.V. Carey, K.F. Hartley, W. Nicholson, P. Taylor, K.P. Tritton, J. Vickerstaff, T. Wilkins, and R. Wood.

I am grateful to B.D. Kelly, R.G. Clowes, H.T. MacGillivray and B.N.G. Guthrie for several useful suggestions on the manuscript and also to R.G. Clowes and D.W.T. Baines for assistance in the preparation of some of the diagrams.

References

[1] Fellgett P.B. (1979), in *Image processing in astronomy*, Eds. G. Sedmak, M. Capaccioli and R.J. Allen, Osservatorio Astronomico di Trieste, pp. 136-154.

[2] Hoffleit D. & Jaschek C. (1982), *The Yale Bright Star Catalogue*, Fourth edition, Yale University Observatory, New Haven, Connecticut.

[3] Jaschek C., (1989), *Data in astronomy*, Cambridge University Press, Cambridge.

[4] Kroenke D. (1977), *Database Processing*, Science Research Associates, Inc., Chicago.

[5] Lawden M.D. (1989), in *ADAM - the Starlink software environment*, Starlink Guide 4.1, Rutherford Appleton Laboratory.

[6] MacGillivray H.T. & Stobie R.S. (1984), *Vistas in Astronomy*, **27**, 433.

19

Database applications in Astronet

Leopoldo Benacchio
Osservatorio Astronomico di Padova
Vicolo dell'Osservatorio 5
I-35122 Padova
Italy

19-1 Foreword

The development and utilization of informational archives and databases started, in the Italian Astronet Project, in the middle of 1983. In that year, in fact, the Special Interest Group (SIG) on Databases and documentation was founded. A small group of astronomers, and some more technical people, met together with a common, painful experience in managing astronomical catalogues with computers.

Two main sets of data were, at that time, substantially available: the SIMBAD system of the Centre de Données astronomiques de Strasbourg (CDS) and some tape collections, as the one of Strasbourg or, a bit later, that of the NSSDC in Greenbelt. The image archives became of interest for the community a few years after the SIG constitution.

Nowadays, some years later, two software packages and the contents of both a relative general database and several local databases represent the work and the effort of the group and of all the people that worked therein during these years. The two archives, the catalogues and images, are managed by two different software packages strongly integrated one in the other: DIRA and DBIMA.

The two systems have been conceived and developed keeping in mind a main goal: to allow the single astronomer to make a free use of original data. An impressive series of small problems regarding formats, documentation, names *etc.* usually hamper astronomers in the world of astronomical catalogues; moreover it requires a certain amount of experience in computer science to develop an *ad hoc* program for each catalogue.

The case for digital images is different: the creation and management of archives is mandatory if an institution is interested in a wider use of images after the proprietary period, in order to avoid losing data forever thereafter.

DIRA's philosophy and design are very simple and proved to be very appreciated by astronomers, namely, to normalize (homogenize), at minimum, astronomical catalogues, to collect satisfactory astronomical documentation on their contents and finally to allow an astronomical approach to the dialogue with the database.

The software foresees two different levels: the general and the proprietary one, respectively called the master and personal level. The main difference, from the user point of view, is that whereas the master level can be accessed only in read mode, in the personal level the user himself is the manager, *i.e.* can create, extend or delete catalogues or archives. The same ideas and philosophy has been ported, with obvious differences, to the image archives and their related software DBIMA.

A description of both systems is given in the next sections. Both software packages are, at the present, strictly dependent on the operating system. A portable version that runs also on Unix systems is under final test.

19–2 DIRA: Background

DIRA is a software structure that, superimposed on the VAX/VMS system, provides users with a simple management of astronomical data files, especially catalogues. This structure can be accessed by the user via a set of VMS-like commands that activate various DCL and Fortran procedures. DIRA allows the creation and management of a catalogues data base, either of general or private use (master and personal data base respectively). The manipulation of files, the extraction of subfiles, by means of user defined criteria, and the iconographic view of the results are all supported.

For special user applications, a Fortran library exists that uses the same philosophy and the same conventions.

The main feature of the DIRA philosophy is that the variables, or physical quantities, contained in the catalogues (Right ascension, Declination, Magnitude, *etc.*) are known to the system by means of conventional codes defined in a vocabulary. This means, in practice, that if one is interested in a list of magnitudes from the Uppsala General Catalogue, one needs only to supply the keywords 'UGC' and 'MAG', besides, of course other more

Database applications in Astronet

trivial specifications on output, *etc.* All commands have a help structure that attempts to clarify possible difficulties when responding to inquiries.

2.1 DIRA main capabilities

DIRA performs primarily the management of data files containing catalogues. The main actions are: the creation of the data base, the updating, the query and data retrieval and the graphic presentation of the output. DIRA commands act on catalogues available at the installation, either on disk or tape and both in the master and/or personal data base, and/or possibly on catalogues (files) that contain the output of a preceding query.

DIRA is able to:

1. list system specifications: *i.e.* the location of catalogue files, their names, reading formats, code of entries in the catalogue record, *etc.*, and also astronomical descriptions of the catalogues and of their contents;
2. create, manage, search, query and retrieve data from the database;
3. provide users with user-oriented utilities for the data display such as sky maps, wedge diagrams, histograms and regression, on both graphic terminal devices and VAX-GPX workstations;
4. provide users with a programmer interface that supports the same functions that are available through the user interface.

2.2 Master and personal data base levels

The distinction between master and personal database is both operational and *philosophical*. This distinction is fundamental for the use of DIRA.

The master DB level is created by the Database Administrator and can be used by all database users in read-only mode. The output of this consultation will produce one or more files that can reside in the personal database area, so the user can access in read/write mode. The personal database (PDB) is a complete subsystem that can, and must be used and managed by a single user. All DIRA system files are copied at the time of the creation of the PDB. Once in the PDB these files can be modified at any time (*e.g.*, the Vocabulary), so possibly customizing the system for a particular research project. The user must, in this case, take care to avoid conflicts between definitions, when accessing also the master level. In the personal database the user can create his own catalogue also through directly editing the files. The two levels can be intermixed in most DIRA commands.

2.3 What is a catalogue in DIRA ?

A catalogue, in DIRA, is composed of two files. The first file (.dat type) contains the data in the format specified at the time of the creation or update of the database. The second file (.doc) contains the associated documentation. The catalogue documentation available originally has been complemented with new entries: number of records, code and format of the variables contained in each record, full explanation of adopted code, *etc.*

These files are normal ASCII files, so that they can be easily updated, modified and manipulated. The use of these files is completely transparent to the user at Master Database level by means of the management commands. At the Personal Database level the user must manage these files, *e.g.* creating documentation files for self-created catalogues.

At the time of writing, the number of available catalogues in DIRA is 122; this figure represents the catalogues in the public level of the database distributed with the software itself on tapes or WORM optical disc. An ISO 9660 CD–ROM is in preparation.

19–3 DIRA structure

3.1 Organization of data

Catalogues in DIRA are organized in tables. Files can have records of variable or fixed length, but the latter are preferred in order to allow users to transform catalogues from variable records structure to fixed length structure and vice versa. A very large blocking size for magnetic tape files is also advised.

We note that in DIRA:

- Every record of a catalogue contains the information relative to one object, and is divided into fields of defined length; it is not possible to use more than one record for the same object;

- The parameters can be represented by integer numbers, floating point numbers or a string of characters. In the case of floating point numbers the use of decimal point is mandatory;
- non homogeneous records within tables (*e.g.* a title) do not exist in a catalogue;
- no restriction exists to parameter formats except for coordinates.

Historically, the coordinates have been defined in a variety of formats

Database applications in Astronet

and it was not possible to use general routines for decoding without losing computer speed. As a tradeoff in between legibility of the catalogue and routine efficiency in system performance we decided to adopt a uniform representation of the coordinates made up of integer digits in the following format:

$$RA = HHMMSSMMM \qquad DEC = \pm DDMMSSMMM$$

where: HH are hours, MM minutes, SS seconds, and MMM thousandth of seconds (Right Ascension), and DD are degrees, MM arcmin, SS arcsec, and MMM thousandth of arcsec (Declination).

3.2 Catalogue names

In DIRA, the catalogues are identified by *logical names* rather than by their physical file names. When possible, the name coincides with the one used by the astronomical community (*e.g.* AGK3, SAO, MCG); in the majority of cases, however, we had to redefine abbreviations and codes. We also tried to make reference to a standard already in use (*e.g.* ADC or CDS naming conventions; Kempt, 1960; Dixon, 1976, *etc.*).

Knowledge of the name of the catalogue is needed to get access to data. When the name with which a catalogue is identified in DIRA is not known to the user, a command can be used to make a search with keys (subject, ADC or CDS identification, ADC or CDS types, band) on the master file of available catalogues (DB_INFO), or to consult the documentation file (DB_DOC).

From an operational point of view the user is not concerned whether the catalogue can be found on disk; the structure provides upon catalogue request either the physical file on disk or the operator will be requested to mount the tape on which the file is found. Obviously, as far as time is concerned, finding the file on disk or on tape will make a lot of difference. To cope with this situation a virtual disk structure has been defined. If at a particular installation there is sufficient space available on disk and a scratch area is available to the user, the database Administrator can establish that the catalogues are copied on disk before they are actually used. This happens anyway in a user-transparent mode. This feature is generally used in very small installation.

3.3 Documentation

For every catalogue there is a documentation file which describes the content from an astronomical point of view. This file is not generally used by application programs; it can be displayed by means of the help routines.

The file contains:

- Logical name of the catalogue;
- relevant bibliography;
- brief description of the catalogue.

The file also describes the parameters contained in the catalogue, the type of representation used for these parameters and the codes with which the parameters are known by the system.

3.4 Vocabulary

The physical parameters of a catalogue are known in DIRA by means of an alphanumeric code. These codes, together with the name of the catalogue are enough information to perform all operations on the data. Therefore, a requirement arose to define a vocabulary of the physical parameters most used in astronomy.

The command DB_INFO allows the retrieval of information from the vocabulary. During the execution of any DIRA command, the help structure assists in showing the codes relative to the selected catalogue, when these are requested.

The database Administrator will provide the definition of new codes when it is necessary, while the user, in the management of the personal database can define his own particular codes (maximum: 8 characters long). In this case, however, the user has to take care not to create conflict between his own designation and that of the system in general, when both are used simultaneously (*e.g.* working on a research on his own catalogue and on one from the system).

3.5 Keys

To facilitate the search of catalogues, some keys have been defined that allow the extraction of particular catalogues in which the user is interested. These keys are used in the command DB_INFO and PDB_INFO and they refer to 4 specified catalogue identifiers.

- *Subject of the catalogue:* to which class of objects the catalogue refers (stars, galaxies, quasars, *etc.*); this key serves also to distinguish between catalogues (bibliography, *etc.*)
- *Identification of the catalogue:* Name of the catalogue in the lists from ADC, CDS and other.
- *Type of catalogue:* in the CDS list convention, allowed types are: astrometric, photometric, spectroscopic, cross identifications, miscella-

neous (see Bull. Inf. CDS, 25).
- *Electromagnetic band to which the catalogue refers:* by entering this key it is possible to distinguish catalogues containing radio data, optical data, and so on.

While working on his own Personal database the user can redefine keys according to his needs.

3.6 Structure

DIRA does not constitute a software environment but can be seen by the user as a set of new commands. DIRA requires, however, its own structure that defines the relationships between the data contained in the database. All the operations carried out in DIRA make reference to a series of files which contain the information relative to catalogues and the definition of the system. Two service files, called Master files, contain all the information regarding the physical organization of the catalogues, while other files define the vocabulary and the keys. The user works with the DIRA structures by means of a series of commands performed at DCL level.

19-4 Personal database

The personal database (PDB) is a structure conceptually identical to the system database, but under the complete responsibility of the single user. This means, in practice, that the database Administrator of a PDB is the user himself.

The structure of the PDB has been implemented in DIRA to satisfy two requirements. First of all, to allow the user to apply on his own files (or tables, or lists) all the command/routines without having to insert such data in the system database. Secondly to make more general the use of DIRA commands. Often the data on which you want to work are a subset of a catalogue. In this case it is wise to first extract the data and put them in the PDB; then you can work with all the commands, in your private area.

All the application commands work indifferently on master (system) or on Personal (private) database; moreover it is possible to insert in the personal area files extracted from master or from PDB. In other words from the PDB you can access the master database.

4.1 PDB scheme

The structure of PDB is identical to the structure of the system master database. The only difference is the location of Master files, Vocabulary files and Key files which are to be found in the user's directory. This allows the user not only to use personal data, but also to modify the vocabulary and search keys.

4.2 Insertion of a catalogue in the PDB

The insertion of a catalogue in the PDB structure of DIRA is carried out in 3 steps.

1. Writing a file with the catalogue data in a compatible format.
2. Writing a documentation file that contains both the full description of the catalogue, and the description of the codes used for the physical variables.
3. Insertion of the information relative to the catalogue into the PDB.

When the catalogue to be inserted in the PDB is a subset of a system catalogue extracted via DIRA commands, the operation (1) and (3) are performed automatically. In this case the documentation file of the extracted catalogue is the original documentation file; when the user works on the documentation, a message warns him that this documentation does not totally correspond to his data.

4.3 Connections between PDB and the DB system

As the reader will have understood, there do not exist in DIRA physical differences between PDB and the DB system, if not for the fact that the files can be found in different locations.

This physical identity is one of the strong points of all the software structure, in which the user can operate the DIRA commands or user written programs indifferently of all the catalogues existing in the center. In describing the commands that work on the catalogues the transportability of data and of information that describes the same data between the two levels of DIRA is guaranteed. Also by using the Fortran routines accessible to the user this possibility has been protected.

Moving from PDB it is still possible to obtain in output catalogues that form part of PDB in a cyclical process that allows successive elaboration.

19-5 Graphic tasks

The main features of the graphic tasks common to all the relative commands are:

- The graphic device independence.
- A user interface based on a command language.
- The use of a loadable log file.

Up to eight catalogues (master and personal together) can be opened at the same time and displayed using different marks and colors.

For each catalogue you can define variables and constraints, while general parameters are defined for all the opened catalogues. The same catalogue can also be opened many times, in this way you can select different sets of objects on the same catalogue.

The graphic tasks work on all the graphic devices (terminal screens, plotters or laser printers) for which a driver is provided. You can work on an old Tek4010 device but you will obtain better results using a color raster terminal or a workstation. The graphic tasks have been developed thinking to an interactive terminal provided with graphic cursor or mouse, with colors capabilities and able to clear graphic objects and screen regions. The window in which the graphic is shown, is called *the working window*.

The alphanumeric terminal used to run tasks can be the same device on which you obtain the graphic output; in this case the software define a window of three rows in the bottom of the screen for the dialogue with the user (we call it *the terminal window*).

Another window, on the right of the graphic screen, shows the principal parameters involved in the operation (we refer to this window as *the parameters window*). The whole graphic screen can be copied on a plotter by means of the hard copy command, or the plotter can be directly addressed as the default output device.

5.1 Graphic device independence

A graphic task in DIRA can work on a variety of graphic terminals by means of different terminal independent graphic libraries: AGL, BGL, GKS. Moreover the management of the alphanumeric terminal is ensured by a terminal independent Window package (TermWindows by François Ochsenbein, ESO). The user must only know few important features of these libraries in order to obtain the best results working with DIRA. TermWindows executes all the operations on the alphanumeric device.

5.2 The User interface

The user interface of the graphic tasks of DIRA is based on about one hundred verbs that execute the graphic operations or define the value of parameters or flags. Since so large a number of commands can give some problems in the first approach, some facilities are provided to help the user in his work. A set of default values are given for typical problems, and the user can change the values when this is required from his applications.

A help structure provides information about commands and values, for instance, '?L' causes the display of the whole list of verbs available in the application; '?' generates a list of the mnemonic codes available. In general, the command 'HELP verb' types the help of the verb.

Typing a verb without parameter(s), when required, causes the system to prompt the user while showing valid value ranges.

The verb 'GO' typed as a first command provides a short way to obtain a guided session.

We have separated verbs into two classes: the *Command verbs* and the *Parameter verbs*. The first one executes operations while the second one defines new values for the program parameters and has no visible effect as long as a command is executed. The log file saves parameters only. For the full description of commands, verbs and parameters available see the graphic tasks description in the DIRA user manual.

5.3 Loadable log file

The whole parameters set used in a graphic session and the task's messages are written in the file DIRALOG.LOG. This is an ASCII file of variable length records that you can print to obtain a permanent log of the operations executed during the graphic session.

19–6 DIRA master database content

The DIRA database continuously grows in terms of *well* documented catalogues. As mentioned before, at the time of writing up to 120 catalogues are present in the master data base distributed with the software. A book collecting all the on line documentation for catalogues is available.

19–7 General characteristic of DBIMA

On the basis of a project carried out originally in in the Astronomical Observatories of Bologna and Padova, software to create and manage archives of astronomical images, has been implemented: DBIMA (DataBase of IMAges).

The archive (or archives, since several image archives can be handled at once) consists of two parts: the data, resident on optical disk, and the logbook, resident on magnetic disk.

The latter holds that part of the archive which can be updated or modified. On the other hand if the whole archive is consolidated and its contents needs no more changes or update, the logbook too can reside on optical disk. The logbook holds the headers of images, thus it is a catalog of these images. Each frame in the catalog is referred by a record in the logbook, that holds the image header, structured by means of descriptive keywords. To manage logbooks, the DIRA software has been adopted. Such a choice presents definite advantages:

- the DIRA software is by now well known to users and diffused inside the Italian astronomical community,
- logbook handling did not require relevant development, since related functions were supplied by DIRA,
- both DIRA and DBIMA realize a uniform approach to data handling for astronomers.

7.1 Looking inside DBIMA

In this section the basics concepts users should know to use DBIMA will be outlined.

7.1.1 Public and Personal Level. Like DIRA, DBIMA can manage public and personal images archives. The distinction between public and personal archives is essential for using DBIMA. The public level holds the archives of general interest, *i.e.* institutional archives; it is created, updated and maintained by the institution through an authorized person, the archivist. Public archives can be accessed by users in the read-only mode. On the other hand the personal level is a complete subsystem inside DBIMA at disposal of users. Inside it, they must perform all the function pertaining the archive management: as a consequence they are responsible for creating and maintaining the archive and for the integrity of related data as well as being the end-users of their own archives.

7.1.2 Input, output and data formats. For input/output operations, DBIMA supports the FITS format: this is in fact the only worldwide adopted system for images transport on tapes (see chapter 25, this book). FITS format is also used for storing data on optical disk, so that inside DBIMA all operations on data refer to a definite and well known data structure.

7.1.3 Logbooks. Since logbooks are catalogues of images headers, they are created by DBIMA in accordance with DIRA file structure. In addition to all catalogues management facilities offered by DIRA, several new tools have been implemented in DBIMA to perform special operations on logbooks. The whole part of the software dealing with images forms the kernel of DBIMA; with this kernel users can perform the following tasks:

- to create the personal archiving environment,
- to create an image archive (logbook),
- to insert images header into a logbook,
- to list the contents of a logbook,
- to read images from tape and store them in a buffer,
- to copy images from magnetic to optical disk, together with their headers,
- to retrieve images from optical to magnetic disks,
- to have a quick look of images.

7.1.4 Hardware and software requirements. DBIMA was developed in VAX/VMS environment, so that, like DIRA, it is dependent on this system. This choice has been made considering that Astronet nodes main computers, at the project start time, were running VMS rather than UNIX operating system. As far as the optical medium is concerned, whatever WORM (Write Once, Read Many times) optical device are suitable and useful, since any access to it is made by means of operating systems calls, ensuring in this way the most efficient access to disk and the highest security level.

19-8 Retrieving data from archives

As logbooks are catalogues of images, users can manage them by means of DIRA commands. In particular all DIRA tasks performing selection criteria, plotting facilities or creating new catalogues at personal level can be useful and help users in handling their own archives.

As often stated above, users can get access to public archives in read-only mode, then they can retrieve from them information and data. DBIMA provides efficient tasks to perform these operations. Following a logical path, users first look for the existence of the datum in the archive, then they may check its quality and finally, if the two previous steps have resulted in success, they may wish to retrieve the datum itself. As this task uses the Images Display Interfaces (IDI) library, only the IDI supported devices are available to perform image quick-look. Since the quick-look is foreseen for a swift inspection of images, it does not offer to users all the standard facilities for image analysis.

In practice, the quick-look task requests the reference number of the target image and displays it with a standard color table taking the image median value from the header and using a simple choice for the dynamic range. To reduce the risk of wrong operations on optical disk, users have a read-only access to data stored on it, so that they cannot directly work on images without having previously copied them into an appropriate area.

In the following references you can find more information on the DIRA philosophy, realization and use; complete documentation, including manuals, is available on request from the author.

References

[1] Nanni M. (1983), XVIth YERAC 13-17 June Onsala (Sweden), *Project of an Astronomical database for Astronet.*
[2] Benacchio L. (1983),in *Astronet 1983*, G. Sedmak (Ed.), p. 109, *Astronet group on database and Documentation: Status report.*
[3] Nanni M. (1983), in *Astronet 1983*, G. Sedmak (Ed.), p. 163, *Software interface to database on tape.*
[4] Benacchio L. (1985), in *Data handling in Astronomy*, Hauck (Ed.), Mem. SAIt, 2, 345, *Astronomical database. Report on some experiences with a new methodology of catalogues management.*
[5] Benacchio L. (1985), in *Data handling in Astronomy*, Hauck (Ed.), Mem. SAIt, 2, 387, *On some problems on the use of astronomical data files. Statement and suggestion for a solution.*
[6] Hunt L., Nanni M. (1985), in *Data handling in Astronomy*, Hauck (Ed.), Mem. SAIt, 2, 565, *A brief description of a Distributed Information Retrieval system for Astronomical files.*
[7] Benacchio L. (1985), in *Astronet 1984*, G. Sedmak (Ed.), Mem. SAIt, *Astronet group on database and Documentation: Status report.*
[8] Nanni M. (1985),in *Astronet 1985*, G. Sedmak (Ed.), Mem. SAIt, *DIRA. A database structure for the management of astronomical catalogues.*
[9] Merighi R., Nanni M. (1985), in *Astronet 1985*, G. Sedmak (Ed.), Mem. SAIt, *The use of DIRA in the ASTRONET link.*
[10] Battistini P., Benacchio L. (1985), in *Astronet 1985*, G. Sedmak (Ed.), Mem. SAIt, *Building a vocabulary for astronomical catalogues use.*
[11] Nanni M. (1987), in *Astronet 1987*, G. Sedmak (Ed.), Mem. SAIt, *The Astronet*

subproject on database.

[12] Battistini P., Merighi R., Nanni M. (1987), in *Astronet 1987*, G. Sedmak (Ed.), Mem. SAIt, *Improvements in DIRA*.

[13] Papers on the new portable version of DIRA will appear in the Proceedings of the Astronet Meeting S.Margherita di Pula, Oct. 1990, to appear in Mem. SAIt 1991.

[14] Battistini P., Benacchio L., Malvasi G. (1987), in *Astronet 1987*, G. Sedmak (Ed.), Mem. SAIt, *Study and implementation of an archive of images on optical discs*.

[15] Papers on the actual version of DBIMA will appear in the Proceedings of the Astronet Meeting S.Margherita di Pula, Oct. 1990, to appear in Mem. SAIt 1991.

20

Database Services at the Canadian Astronomy Data Centre

A.W. Woodsworth
Canadian Astronomy Data Centre
Dominion Astrophysical Observatory
National Research Council of Canada
5071 West Saanich Rd.
Victoria B.C., Canada V8X 4M6

20-1 Context

1.1 History

The Canadian Astronomy Data Centre (CADC) was initiated by the Dominion Astrophysical Observatory (DAO) to provide access to the HST data archives for Canadian scientists and to promote involvement in space astronomy research. A secondary goal is to provide a small centre of expertise in astronomical computing within the Canadian community. These goals were identified by the Canadian Astronomical Society and its subcommittee on Space Astronomy in 1984. Canada is not a partner in the HST mission, nevertheless NASA has agreed to provide the CADC with a copy of the non-proprietary data. CADC must pay for the incremental costs of producing the copy, which are a tiny fraction of the actual costs of the HST mission. These costs are being reimbursed by the Canadian Space Agency.

The original expectation was that CADC would simply use the Data Management Facility (DMF) software which was being developed by others (ST–ECF and STScI) so startup would be relatively easy. However, the CADC's geographical location on the western edge of a very large country has led to an emphasis on services for remote users and a strong involvement in the development of a national research communications network (CA*net). Additionally, CADC has become involved in developing various components of the DMF software package, which was not complete at the time of writing.

Other data services from space- and ground-based observatories are also provided or planned, and these are described below.

1.2 Description

The Canadian Astronomy Data Centre is located within the Dominion Astrophysical Observatory (DAO), which is near Victoria, British Columbia in a scenic, park-like setting. The DAO is a component of the Herzberg Institute of Astrophysics, which itself is a Division of the National Research Council of Canada. Staffing of the CADC was held back following the Challenger disaster, and the subsequent postponement of the HST mission. However, we expect to achieve a level of six people during 1990. The expected staff level of six includes two astronomers, two programmer-analysts, an archivist and a manager (the author). In addition, the CADC staff interact with a number of scientists at DAO involved in HST and other space-based research. Policy guidance is provided by a committee known as the Joint Subcommittee on Space Astronomy, which reports to the Canadian Space Agency, the Canadian Astronomical Society and indirectly to the National Research Council.

The Canadian Astronomy Data Centre is involved in joint software development projects with the groups at STScI in Baltimore and the ST-ECF in Garching. The initial work was done using microVAX computers and a Britton-Lee Intelligent Database Machine, but recently efforts have been switched to Sun SPARCstations which support the Sybase relational database management software.

1.3 Personnel

The following personnel are currently part of the CADC project: Daniel Durand, Dennis Crabtree and Stephen Morris are astronomers with strong software backgrounds who will be responsible for developing expertise in dealing with data from specific HST instruments. Gerald Justice is a computer programmer-analyst, Wes Fisher is a technologist who will be acting as the archivist-librarian and providing user assistance, and Andrew Woodsworth is an astronomer who is acting as the CADC manager. All of the astronomers are expected to undertake research programs as well.

20-2 Databases and Services

2.1 HST Services

The principal services are those relating to HST data products, and these are described elswhere in this volume (chapter 6, page 47). However, in summary we support:

- STARCAT, which provides the screen-based user interface to the HST archives.
- Data delivery from the CADC copy of the HST archive, using Data Management Facility as requested through the STARCAT interface.
- Remote access to the Guide Star Catalog, through the Space Telescope Science Data Analysis System (STSDAS) package of IRAF . This catalog is on two CD–ROMs which are permanently mounted.
- Access to Calibration Database—this will be useful to Guest Observers as well as Archival Researchers who wish to recalibrate their data. Recalibration software is available within STSDAS.
- HST data analysis software, including the STSDAS package provided by STScI, and also the instrument team software for the Goddard High Resolution Spectrograph, which is written using the proprietary IDL software. Additionally, the popular DAOPHOT package is being modified at DAO by its developer (Dr. Peter Stetson) for use with Wide Field/Planetary Camera images.

2.2 Other Space-based data services

CADC has a complete copy of the IRAS data (including all the HCON images) as well as the IUE low- and high-resolution spectra. We also have installed the IUE Vilspa ULDA product (Uniform Low Dispersion Archive) but staff shortages have inhibited our ability to support this. Nonetheless, we expect the ULDA to become more valuable in the future, for planning HST and Lyman observations.

2.3 Ground-based data services

The STARCAT system includes a number of popular catalogs of ground-based data, and CADC has added a few more catalogs (such as the NASA Infrared survey, Gezari 1987).

One of the most popular services before HST launch has been the provision of access to SIMBAD. CADC provided this even before it was available to US astronomers, to make access easier and to allow dialled access using a modem for astronomers working at home or at telescopes. The service is currently provided without charge, as CADC pays for the X.25 access charges to France as well as the SIMBAD cpu charges. In the near future, SIMBAD access will be available over the Internet. At that time, it might be simpler to simply provide Canadian users with the CADC account password and let them connect directly themselves, rather than through a CADC host computer.

20–3 Access Mode

3.1 X.25 Access

During the past two years, access has primarily been through the public packet switching network (X.25), known as Datapac in Canada. Users can connect to Datapac from all major Canadian cities by dialled access at speeds up to 2400 baud, or through a private Packet Assembler-Deassembler (PAD) at speeds up to 19.6 Kbaud. The CADC has been paying the access charges to encourage use of the services, although this policy is reviewed from time to time.

However, the speeds available and the type of access possible over Datapac will be too limiting for the HST era. For instance, implementations of TCP/IP over X.25 are not common, so most users only have a terminal access capability over X.25, and no file transfer methods. This may change as higher level ISO/OSI applications are implemented within Unix, VMS and other popular operating systems (see also chapter 23, this book).

3.2 CA*net

Because of the importance of good high-speed network access to the CADC, we have taken a lead role in the development of a national computer network. CA*net is a national backbone network, which interconnects regional networks in all parts of Canada except the far north. This network is loosely modelled on the highly successful NSFnet in the US, and in fact connects to NSFnet as a peer network.

CA*net supports the TCP/IP protocols, commonly used on Unix systems but available for all popular computers from PC's to IBM mainframes. In the future, it will also support OSI. The intial backbone speeds are 56 Kbit/s, and redundant links are provided which also provide parallel bandwidth.

20–4 User Profile

The users of CADC services are astronomers across Canada. They are, for the most part, university faculty members and their students and post-doctoral fellows. Other astronomers are located in the laboratories and observatories of NRC's Herzberg Institute of Astrophysics, including DAO.

Some members of the Canadian astronomical community are actively involved as Guest Observers and Guaranteed Time Observers on HST and as Guest Observers on the IUE satellites. These users are quite sophisticated and generally know exactly what they need. However the majority of Canadian astronomers are new to space astronomy techniques. Initially, we expect that they will use the CADC archives on an occasional basis. For instance, they may request Wide Field/Planetary Camera images in much the same way they now use the Palomar Sky Survey plates. These individuals will need much more direct assistance in their use of the available software, and they may need direct assistance in their selection from the available data. One of our goals is to help the latter class of users to become sophisticated practitioners of space astronomy data analysis.

20–5 Future Plans

In the first instance, the future of the CADC is intimately linked to the future of the HST. The reduced performance arising from the optical focus problems will probably slow the growth of the HST archive during the first two to three years of operation, as well as redirecting the emphasis toward spectral, rather than imaging data.

However, the Canadian community is increasingly aware of the burgeoning field of archival database research in astronomy. Many roles *could* be played by CADC, depending on community priorities and the interactions between these priorities and CADC's personnel and financial levels. Some examples follow:

- The CADC has also agreed to maintain an archive of data obtained from the Canada-France-Hawaii Telescope, located on Mauna Kea. This will be done using the same hardware and software combination as is used for HST data, so that both will be accessed in the same way. Initial emphasis will be on imaging data. Unfortunately, manpower shortages at CADC and CFHT have prevented implementation to date.
- Canada will be a partner in the Lyman ultraviolet satellite, to be launched in 1997, and the CADC is expected to provide data archiving services for this mission.

- Canada is also negotiating with Britain and the United States regarding a collaborative project to construct two 8-m telescopes, and if this project goes ahead with Canadian participation, the CADC is again expected to provide archiving services.
- In addition, CADC could install databases from other satellites (e.g. Einstein, ROSAT) if demand and resources levels warrant this.
- Finally, the CADC would like to undertake a key research project using the HST archives. This might be some sort of classification project, but this has not yet been decided because staffing levels are still too low to undertake this sort of project.

5.1 Impact of DADS

NASA has contracted with Ford Aerospace to develop a follow-on system for the distribution of HST data. This system, known as the Data Archive and Distribution System (DADS), is planned to be available in the 1992-93 timeframe. It will allow remote access to the database as well as distribution of archival copies on optical disk, like the current *Data Management Facility* system. While the current specifications for the DADS system were not developed to support access by international users, this could be technically quite feasible given the rapid development of international high speed data networks.

It may turn out, with the development of DADS and the higher level Astrophysics Data SystemADS, that the rationale for a separate Canadian copy of the various archival databases diminishes. If this were the case, the CADC could concentrate on user assistance rather than database maintenance. However, given the history of major developmental systems, it seems likely that we will be providing HST services via *Data Management Facility* for the foreseeable future.

References

[1] Gezari D. Y., Schmitz M., Mead J. (1987), *Catalog of Infrared Observations*, NASA Reference Publication **1196**.

[2] Stetson P. B. (1987), *Publications of the Astronomical Society of the Pacific*, **99**, 191.

21

Astronomical bibliography from commercial databases

Joyce M. Watson

Harvard-Smithsonian Center for Astrophysics
60 Garden Street
Cambridge, MA 02138, USA.

21-1 Introduction

There are now many databases that concentrate on astronomical data from specific missions, such as the IUE, IRAS and EXOSAT, to name three, which can be accessed via the facilities at which they are archived and made available (see related chapters, this book). We have come a very long way in the last few years in improving the access to these data.

In my first surveys, published in 1983 and 1988 (Rey 1983; 1988), it appeared that tracking down much information depended upon serendipity, good detective work, and finding out whom to call with what question. This situation has actually changed very little. We are now, we hope, moving rapidly, but by no means painlessly, towards a comprehensive source of all astronomical data, including the bibliographical, and made available as expeditiously as possible to all interested researchers, through systems such as the emerging NASA Astronomical Data System, or ESA's European Space Information System (see chapters 15 and 14, this book).

The greatest step forward, undoubtedly, was the advent of SIMBAD (see chapter 9) as a resource accessible to the international astronomical community. Here, at last, searches both by stars and stellar objects, by coordinates, or by other specific parameters, could be set up. The bibliography, with its wide coverage, is undoubtedly the finest available anywhere, and one can rely upon both its relevance and its timeliness.

The problems of searching by subject, however, still exist. It is possible to perform a good literature search, depending upon one's needs, in the databases described below. It must always be remembered that these sources, excellent though they may be, are multidisciplinary in character, and, perforce, attempt to cover *all* possible aspects of the physical sciences, of which astronomy is but one small part. The producers of these commercial databases, of course, are faced with the necessity not only of satisfying as many diverse disciplines as possible, but also of showing a profit. The cost of producing a database is high, as is the cost of searching it.

Thus, there exist no commercial databases specifically dedicated to astronomy and astrophysics. It is necessary to improvise and compare results. No mention is made here of search costs, as this is an ever-changing (increasing) element. At present, STN PHYS, INSPEC, and SCISEARCH are costly but available —the NASA database is inexpensive, but not necessarily accessible. THE AEROSPACE DATABASE is also of limited availability and high cost. No attempt has been made in describing the databases below in most cases to list the vendors through which access may be obtained, as many of them are available from several vendors.

Our hopes of seeing the entire series of *Astronomy and Astrophysics Abstracts* as a separate and complete database now appear unlikely to be fulfilled. We must therefore sincerely hope that the agreement between the owners of the PHYS database and the owners of the *Abstracts* will be duly honored, and that at least in the future we can rely on the same complete coverage of the *Abstract* in PHYS as we must now obtain manually from the bound volumes.

On-line retrieval systems, in general, consist of two components: the data (called databases) and the retrieval infrastructure (called hosts). Usually, companies running the retrieval systems (called vendors) are independent of the database producers and load into their systems a large number of databases from different sources. In the next sections, descriptions of astronomically useful databases are given (in alphabetical order), followed by a list of the major scientific database vendors.

21-2 Databases

2.1 Astronomy and Astrophysics Monthly Index

Print equivalent : same title
Produced and owned by : Olivetree Associates,
Sierra Madre, CA 91025, USA.
Updated : Monthly

This compilation is a privately produced index covering commercially available journals, observatory publications, preprints, and conference proceedings. It is a quick announcement service which attempts to bridge the gap between the semi-annual issues of *Astronomy and Astrophysics Abstracts*.

It is available in a printed version or as tapes distributed monthly for mounting on inhouse computers for local searching; hand-tailored searches can also be requested. As the accent is on timeliness, no abstracts and no annual index are produced. The average monthly issue contains in excess of 1000 new article titles.

2.2 CONF (CONFerences in Energy, Physics and Mathematics)

Print equivalent :	None
Produced and owned by :	STN International
	c/o Fachinformationszentrum
	Energie-Physik-Mathematik GmbH (FIZ)
	Postfach 2465
	D-7500 Karlsruhe 1, Fed. Rep. of Germany
Contact (Europe) :	Telephone : (+49) 7249/808-555
	Telex : 17724710
	Facsimile : (+49) 7247/808-666
Contact (U.S.) :	STN-Columbus
	c/o Chemical Abstracts Service
	American Chemical Society
	2540 Olentangy River Road
	P.O. Box 2228
	Columbus, OH 43202, USA
	Telephone: 614/421-3600
	800/848-6533
	Telex : 6842086 chmab
Timespan covered :	1976 to date
Updated :	Weekly
Items added :	About 100 conference records and 100 records per week

This database covers information on conferences in the above subjects, which include the space sciences, astronomy and astrophysics. Conferences are tracked from first announcement, press releases, calls for papers, contacts, etc., and notices are included sometimes years in advance of meeting dates. However, CONF does *not* include listing of individual papers presented at the meetings.

2.3 Conference Papers Index (CPI)

Print equivalent :	Same title
Produced and owned by :	Database Services
	Cambridge Scientific Abstracts
	7200 Wisconsin Avenue

Contact :	Bethesda, MD 20814, USA
	Telephone : 800/843-7751
Timespan covered :	1973 to present
Updated :	Bimonthly
Items added :	8 000 per update

CPI covers about 50,000 papers of approximately 150 scientific meetings per year, worldwide. It covers life sciences, engineering, and physical sciences.

As distinct from the CONF file described above, CPI does *not* list specific meetings individually, although this information can be extracted. It can be searched by authors, by subject headings, title of article, in conjunction with date, conference location, source, etc. It is useful in identifying obscure references to conference papers, although, as is always the case with multidisciplinary publications and databases, as explained above, the coverage of astronomy can never be regarded as comprehensive.

2.4 Current Contents Search

Print equivalent :	CURRENT CONTENTS: PHYSICAL, CHEMICAL AND EARTH SCIENCES
	(Other editions cover different disciplines.)
Produced and owned by :	Institute for Scientific Information (ISI)
	3501 Market Street
	Philadelphia, PA 19104, USA
Contact (U.S.) :	Telephone : 215/386-0100
Contact (Europe) :	Institute for Scientific Information
	132 High Street
	Uxbridge, Middlesex UB8 1DP, England
	Telephone : +44 895-70016
	Telex : 9933693 UKISI
Timespan covered :	1988 to present
Updated :	Weekly (15,000 records per update)

This weekly publication of the Institute for Scientific Information (ISI), now available online, is, literally, a collection of reproduced tables of contents of journals, which often appear here before the journal issue itself is available. The coverage of astronomical and related journals is good, and the database is also an excellent source of items published in obscure or marginal journals that might otherwise be overlooked. Approximately 6,500 journals in every discipline within the sciences, social sciences, arts and humanities from the different editions of *Current Contents* are covered in the database.

Most vendors offer the facility for setting up "saved searches," in many commercial databases, and it is possible to set up a personal profile. This feature is particularly useful in Current Contents Search, as profiles of authors, journal titles, etc. can be input, saved, and accessed regularly to obtain updated information.

2.5   INSPEC (Information Services in Physics, Electrotechnology, Computers and Control.)

Print equivalent :	*SCIENCE ABSTRACTS,* Section A: *PHYSICS ABSTRACTS,* Section B: *ELECTRICAL AND ELECTRONICS ABSTRACTS,* and Section C: *COMPUTER AND CONTROL ABSTRACTS*
Produced and owned by :	The Institution of Electrical Engineers, Ltd. (IEE) Station House Nightingale Road Hitchin, Herts., SG5 1RJ, England.
Contact (Europe) :	Marketing Manager Telephone : Telex : 825962 IEE G Facsimile : 0462 59122
Marketed in U.S. by :	(Institute of Electrical and Electronic Engineers, Inc.) INSPEC IEEE Service Center 445 Hoes Lane Piscataway, NJ 08855-1331, USA.
Contact (U.S.) :	Telephone : 201/562-5549 Telex : 833233 IEEE PWAY Facsimile : 201/981-0027
Timespan covered :	1969 to present
Updated :	Semimonthly, corresponding to issues of printed version
Items added per year :	149,000

INSPEC covers the whole field of physics, and includes the core journals of astronomy and astrophysics, and good results can usually be obtained. It is particularly valuable in searching peripheral topics, such as atomic and molecular physics, gravitation, cosmology, etc. Methods exist for searching numerical data which are difficult to obtain elsewhere, e.g., limitation by wavelength, etc. It is also valuable in searches of space engineering and computer techniques, as a result of the extremely wide coverage of topics. (It is possible to specify which section should be searched.)

2.6   NASA/RECON and The Aerospace Database

Print equivalents :	*NASA SCIENTIFIC AND TECHNICAL*

	AEROSPACE REPORTS (STAR)
	INTERNATIONAL AEROSPACE ABSTRACTS (IAA)
Produced and owned by :	U.S. National Aeronautics and Astronautics Space Administration (NASA)
	NASA Scientific and Technical Information Facility
	P.O. Box 8757
	Baltimore/Washington International Airport, Maryland 21240, USA
Contact (U.S.) :	Telephone : 301/859-5300
	301/621-0300
Timespan covered :	1963 to present
Updated :	Weekly
Items added per year :	Not specified – file size exceeds 2,000,000 records

The NASA/RECON database is the earliest and most comprehensive of databases encompassing the entire field of space and space-related science, including astronomy, and dates back to 1962. The printed titles are complementary, both covering government-funded research. *STAR* covers unpublished report literature, including research funded not only by NASA but also by other U.S. government agencies, foreign material, translations, contributed papers emanating from NASA-sponsored meetings, and other miscellaneous material. *IAA* similarly covers the open literature, journal articles, published conference proceedings, etc.

Access to the NASA/RECON database was originally limited to NASA-related facilities only, but was opened to academic and other interested institutions in the U.S. in the early 1980s. A version is also available via ESA–IRS in Europe. In 1985, another version was made available as the AEROSPACE DATABASE through Dialog Information Services, Inc., but accessible in the United States only. This source includes all material from STAR and IAA, but omits certain categories of material included in the main RECON. In June 1990, the EUROPEAN AEROSPACE DATABASE started its operations as a focal point for technical reports from the aerospace industry in Europe. This database is available only through ESA/IRS and includes only material originated in ESA member states.

Unfortunately, these limitations tend to deprive users of access to the database considered by the author to be the most comprehensive source of astronomical information. For subject and author searches, in her experience, the NASA database has consistently scored a higher number of hits in comparison searches than any of the other sources listed here.

2.7 PHYS

Print equivalent :	*PHYSICS BRIEFS/PHYSIKALISCHE BERICHTE*
Produced and owned by :	Fachinformationszentrum
	Energie - Physik - Mathematik GmbH (FIZ)
	Postfach 2465
	D-7500 Karlsruhe 1, Deutschland
Contact (Europe) :	Telephone : (+49) 7249/808-555
	Telex : 17724710+
	Facsimile : (+49) 7247/808-666
Marketed in U.S. by :	American Institute of Physics
	335 East 45th Street
	New York, NY 10017, USA
Contact (U.S.) :	Marketing Supervisor
	Telephone : 212/661-9261
	800/AIP-OHYS
	Telex : 960983 AMINSTPHYS NYK
	Facsimile : 212/661-2036
	212/949-0473
Timespan covered :	1979 to present
Updated :	Semimonthly
Items added per year :	130,000

This file is available only from STN International in Karlsruhe. In the United States, it can be accessed via the STN gateway in Columbus, Ohio. It contains abstracts for journals, books, patents, reports, theses, and conference papers published since 1979 in all major languages, and on all aspects of physics.

During the last few years, a concerted effort was made to increase the coverage of astronomy and astrophysics in PHYS. Not only has the coverage been increased, but an agreement was recently reached with the Astronomisches Rechen-Institut in Heidelberg to incorporate tapes from *Astronomy and Astrophysics Abstracts*. These are not to be incorporated retrospectively, unfortunately, but PHYS should eventually become the database of choice in astronomical searches.

New features include the addition of object designations in the title field, thus adding to its flexibility, but it is still necessary to use SIMBAD for comprehensive object searches. PHYS provides good coverage in related areas of physics, although it is necessary to revert to the use of INSPEC for citations before 1979. Updating is very prompt.

2.8 SCISEARCH

Print equivalent :	*SCIENCE CITATION INDEX*, plus additional records from *CURRENT CONTENTS*
Produced and owned by :	Institute for Scientific Information (ISI)
	3501 Market Street
	Philadelphia, PA 19104, USA
Contact (U.S.) :	Telephone : 215/386-0100
Contact (Europe) :	Institute for Scientific Information
	132 High Street
	Uxbridge, Middlesex UB8 1DP, England
	Telephone : +44 895-70016
	Telex : 9933693 UKISI
Timespan covered :	1974 to present
Updated :	Weekly
File size :	About 12.5 million records.

This is the online version of the *Science Citation Index*, produced by the Institute for Scientific Information (ISI). Input to the database includes some timely or pre-publication material from *Current Contents* (see above). SCISEARCH is a multidisciplinary index for all fields of science and technology and contains well in excess of 12 million entries, which date back to 1974.

The uses of SCISEARCH for our purposes vary from other databases described here. No abstracts are included, so subject searches must be performed by title, author, journal. It covers a wide selection of astronomical journals, and searching by author particularly produces good results.

The exclusive and most valuable feature of SCISEARCH, however, lies in the indexing of citations. All references to previous works cited by an author in footnotes or the bibliography of an article are indexed. It is possible to input a reference, for example, published twenty years ago, and to generate a list of all subsequent papers which make reference to it. Thus, the direction taken by a research topic from the article reporting the original work can be tracked, together with the scientists who have referred to it in their own publications, and the institutions where they have pursued their research. There is no time limitation to cited references; works are included as they are cited, in some cases back to the early years of this century, or even before.

Other useful applications of this feature are obvious. A quick check can be made of the publications of a scientist, and an assessment made of the quality of his or her research by the number of times the works were cited, where, when, and by whom. Thus bibliographies can be generated, and

2.9 NTIS

Print equivalent :	GOVERNMENT REPORTS ANNOUNCEMENTS
Produced and owned by :	National Technical Information Service
	U.S. Department of Commerce
	5285 Port Royal Road
	Springfield, VA 22161, USA.
Contact (U.S.) :	Telephone : 703/487-4600
	Telex : 89-9405 or 64617
	Facsimile : 703/487-4630
Contact (U.K.) :	Microinfo Limited
	NTIS Division
	P.O. Box 3
	Alton, Hants GU34 2PG, England
	Telephone : (+44) 420-86848
	Facsimile : (+44) 420-89889
Contact (France) :	World Data
	Mr. Boris Prassaloff
	B.P. 68
	75060 Paris 2, France
	Telephone : (+33) 1-4508-8566
	Facsimile : (+33) 1-4278-1472
Contact (Italy) :	Diffusione Edizioni Anglo-Americane
	Librerie Internationale
	28 Via Lima
	00192 Roma, Italy
	Telephone : (+39) 06 855-1441
	Facsimile : (+39) 06 854-3228
Timespan covered :	1964 to present
Updated :	Biweekly (about 5,000 records/month)
File size :	In excess of 1.5 million records

NTIS is one of the largest and most comprehensive databases, available worldwide through local branches of the large vendors listed below. Its principle coverage is the report literature on research supported by the Department of Defense, the U.S. Air Force and Navy, and other government agencies, including many NASA reports.

Whenever performing searches in astronomy, or most other subject areas for that matter, it is good practice to include a run of NTIS. NASA is by no means the only U.S. government agency engaged in research in astronomy, and unexpected gold mines are quite often to be found in NTIS.

NTIS is also an excellent supplier of government-published material, which can be ordered online by a wide variety of systems in many countries.

21–3 Vendors

A company that offers access to a number of online databases is known as a vendor. By registering with the local office of any vendor, you will obtain access to the databases of your choice. These cover every conceivable subject from international commerce to pure science. You will use a modem or other means to connect to the central facility, choose which file you wish to use, and be charged according to the time you spent on your search, and the database to which you were connected.

Five of the best-known vendors are listed below. They are world-wide in scope, and offer introductory and advanced training programs, both in their systems and in specific databases or subjects, to enable you to become proficient and obtain the best results from your searches.

3.1 BRS (Bibliographic Retrieval Service) and ORBIT Search Service

are both operated by the following:

U.S.A.	Maxwell Online, Inc.
	8000 Westpark Drive
	Maclean, VA 22102, USA
Contact (BRS) :	Telephone : 703/442-0900
	800/289-4277
Contact (ORBIT) :	Telephone : 800/456-7248
	800/45-ORBIT
	Facsimile : 703/893 4632
ENGLAND	Maxwell Online, Ltd.
	Achilles House
	Western Avenue
	London, W3 OUA, England
Contact (both) :	Telephone : (+44) 01 992-3456
	Facsimile : (+44) 081 993-7335

Maxwell Online, Inc., also has offices in Australia, Korea, Japan, etc.

3.2 Dialog

Dialog Information Services, Inc.
3460 Hillview Avenue

	Palo Alto, CA 94304, U.S.A.
Contact :	Telephone : 800/3-DIALOG

Dialog is by far the largest database vendor, and has offices in most countries of the world. Training programs are offered in the larger urban centres, in the language of the country. Dialog offers access to approximately 500 different databases, including many of those described above.

3.3 STN International (The Scientific and Technical Information Network)

Germany	STN-Karlsruhe
	c/o Fachinformationszentrum
	Energie, Physik, Mathematik, GmbH
	Postfach 2465
	D-7500 Karlsruhe 1
	Deutschland
Contact :	Telephone : (+49) 7247/82-4566
	Telex : 7826487 fize d
U.S.A.	STN-Columbus
	C/o Chemical Abstracts Service
	American Chemical Society
	2540 Olentangy River Road
	P.O. Box 2228
	Columbus, OH 43202, USA
Contact :	Telephone : 614/421-3600
	800/848-6533
	Telex : 6842086 chmab

There are also STN facilities in Japan and Australia.

3.4 ESA/IRS (Information Retrieval Service)

Operated at the European Space Research Institute (ESRIN), this system includes up to 130 databases covering a wide range of science domains.

	ESA/IRS
	Via Galileo Galilei
	Cassella Postale 64
	I-00044 Frascati
	Italy
Contact :	Telephone : (+39) 6 941 801
	Telex : 610637 esrin i
	Telefax : (+39) 6 941 80361

References

[1] Rey J. M. (1983), "Information sources and services in astronomy, astrophysics, and related space sciences", Cambridge, MA, Smithsonian Astrophysical Observatory, *Smithsonian Institution Libraries Research Guide*, **2**.

[2] Rey-Watson J. M. (1988), in *Astronomy from Large Databases*, F. Murtagh & A. Heck (Eds.), ESO Proceedings **28**, 453.

22

Astronomical Directories

André HECK
Observatoire de Strasbourg
11, rue de l'Université
F-67000 Strasbourg
France

22–1 Introduction

Belgians seem definitely to have a special relationship with astronomical directories, as we shall see in this paper. Such systematic compilations go back as far as the beginning of the century and include nowadays, not only typical astronomical data, but also entries of general interest, reflecting how much diverse astronomy has become and how many fields are connected to our activities.

Astronomical directories have been up to now delivered as compilations on paper, with an indexing flexibility constantly increased by the improvements of the desktop publishing. Today plans are being finalized to make available some of the directories as on–line databases, taking advantage of the most recent advances in networking and database interconnectability.

22–2 First compilations

The first astronomical directory of modern times has been produced at the Royal Observatory of Belgium (ROB), back at the very beginning of the century (Stroobant *et al.*, 1907). In the foreword, the practical and most useful characters of the compilation are stressed. The presentation of each entry is in French and is literary in the sense that each type of data is fully described. For an observatory, one gets the geographical coordinates (with the reference of the source), a personnel list (with the corresponding titles and positions) and a historical note as well as a description of the instruments and activities. The observatories are listed alphabetically on the location name.

Astronomical societies are also included with details on their foundation, aims, activities, publications and constitution of the board. In addition, a few journals are mentioned with indications on their foundation, editors, publication frequencies and prices as well as on their contents and volume status. Finally, indices of names of locations are given together with a few pages where societies and journals are sorted by countries. A map showing the distribution of institutions at that time has been reproduced here in Fig. 22–1.

Updates and complements have been produced later on (Stroobant *et al.*, 1931; Rigaux, 1959 & 1961). According to a report of IAU Commission 5 (Pecker, 1979), it did not seem that the work carried out at the ROB was going to know further updates.

22–3 The virus strikes back

Although sharing the same nationality as the authors mentioned in the previous section and unaware at that time of their work, we were led to compile astronomical directories through quite a different approach.

If, one day, you decide to publish a book for astronomers, amateur or professional ones, you have, of course, to make sure that your publisher has all the elements necessary to sell your masterpiece to the astronomical community and, possibly, to the outside world. Thus, after putting together an astronomical photographic atlas (Heck & Manfroid, 1977), we started gathering lists of essentially amateur organizations around the world. They were indeed the prime marketing targets since they could have bought the book for their libraries and/or let it know to their members. Lists of journals were also set up for taking advantage of the reviewing system.

Then, it was realized that such lists could have an intrinsic interest and that it was worthwhile to investigate how their publication would be received. Normally they would improve or facilitate the establishment of national and international relationships in amateur astronomy as well as providing professional institutions with lists of groups they could approach for *e.g.* complementary observations. They had also an historical value since they were providing a picture of the world of amateur astronomy in those years.

Thus came to light the 1978, 1979, 1981, 1982 and 1984 versions of IDAAS under its original title *International Directory of Amateur Astronom-*

Astronomical Directories 213

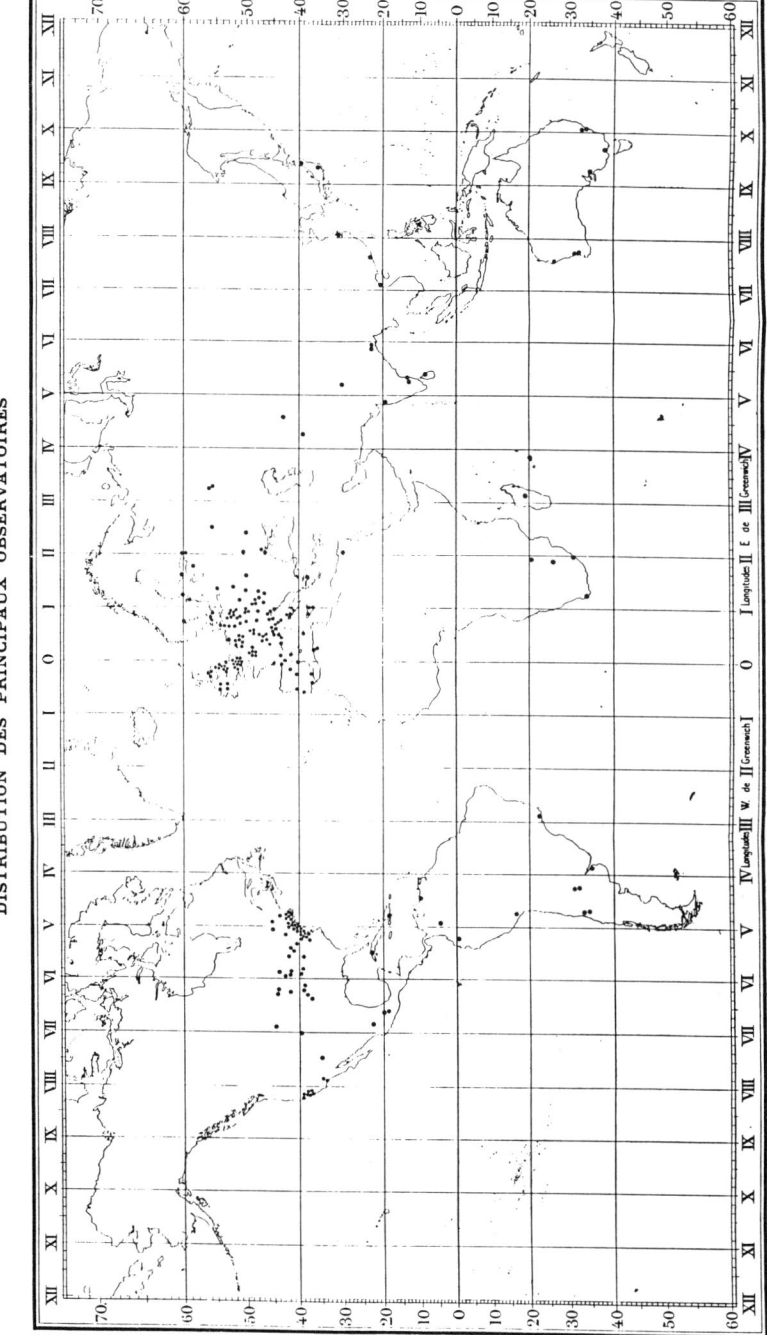

Figure 22-1: World distribution of the main observatories after Stroobant *et al.* (1907)

Year of edition	Number of countries	Number of entries	Number of pages
1978	27	≈600	112
1979	56	>1,200	290
1981	51	>1,100	304
1982	54	>1,200	308
1984	56	>1,200	282
1986	58	>1,100	270
1988	64	>1,700	522
1990	90	>3,200	724

Table 22–1: Statistics on IDAAS editions

Year of edition	Number of countries	Number of entries	Number of pages
1987	68	>1,500	280
1989	80	>2,700	498
1990	90	>3,500	666

Table 22–2: Statistics on IDPAI editions

ical Societies. Later on, the 'IDAAS' meaning was changed to *International Directory of Astronomical Associations and Societies* because more and more professional societies were included and the word *amateur* was no more appropriate in the title. Three additional editions (1986, 1988 and 1990) were then produced (see for instance Heck, 1989a).

Along the years, more and more professional institutions showed an interest in the successive IDAAS editions. Therefore a list of institutions was also compiled for advertising IDAAS and, since the list was existing, why not to publish it too? Hence came to light the other directory IDPAI, standing for *International Directory of Professional Astronomical Institutions.* Three editions have been produced: 1987, 1989 and 1990 (see for instance Heck, 1989b).

Tables 22–1 & 22–2 give a brief description of the successive IDAAS and IDPAI editions. The first ones were produced with the most valuable collaboration of Jean Manfroid.

From the start, these directories received an enthusiastic welcome and, along the years, numerous letters of support had encouraged us to continue the work and to broaden the scope of the directories. Therefore the last editions were bearing the sub–title *together with items of general interest.* In the following section, we shall describe in more detail what were the contents of these directories that will be merged from now on into a single one entitled *Astronomy, Space Sciences and Related Organizations of the World* (ASpScROW) .

22–4 Present situation

ASpScROW gathers together all practical data available on associations, societies, scientific committees, agencies, companies, institutions, observatories, universities, etc., more generally organizations, involved in astronomy and/or space sciences. Related fields such as aeronomy, astronautics, atmospheric sciences, geodesy, geophysics, meteorology, and so on, are also covered when justified.

Many other entries are also included: academies, bibliographical services, data centres, dealers, distributors, journals, funding organizations, IAU–adhering organizations, manufacturers, meteorological services, national norms and standards institutes, parent associations and societies, publishers, software producers, and so on.

The information is given in an uncoded way for easy and direct use. For each entry, all practical data available are listed: city, postal and electronic-mail addresses; telephone, telex and telefax numbers; foundation years; numbers of members or staff; main activities; titles, frequencies, ISS–Numbers and circulations of periodicals produced; names and geographical coordinates of observing sites; names of planetariums; awards, prizes or distinctions granted; and so on.

The entries are listed alphabetically in each country. At the end of the volumes, an exhaustive index gives a breakdown not only by different designations and acronyms, but also by location and major terms in names. Categorial and thematic subindices are also provided as well as statistics on the contents (numbers of entries per country, memberships, years of foundation) and a list of telephone, telefax and telex national codes.

At the time of writing, ASpScROW gathers already about 6000 entries from about 100 countries. This will lead to a book of more than 1200 pages most likely published in two volumes (Heck, 1991).

A list of acronyms was also included in the 1990 editions of IDAAS and IDPAI, but it has become so voluminous while compiling ASpScROW 1991 (presently about 30,000 entries), that it became more appropriate to provide it as a separate, nevertheless complementary, publication (Heck, 1990). It is printed at request as it is continuously expanded and kept up-to-date.

Figures 22–2 to 22–4 display the distribution of the astronomical and related professional institutions listed in the ASpScROW version in preparation and for which coordinates have been provided to us. Observing and receiving stations depending from these institutions have also been included. About 700 coordinate sets were provided, with, of course, some redundancy. Some points are also located very close to each other. Hence the number of points that can be counted on the maps is significantly lower.

Again, it is necessary to insist on the fact that these maps are representative only of the data provided on the questionnaires returned to us and do not pretend to give an exhaustive picture.

Nevertheless, some obvious conclusions can be drawn from the maps themselves and also by comparing them with the older distribution reproduced in Fig. 22–1. While points have multiplied about everywhere (and particularly in Europe and North America), third–world countries and especially Africa are definitely less well represented.

Moreover, the multiplication of points between Fig. 22–1 on one hand and Figs. 22–2 to 22–4 on the other hand should be weighted by the fact that we had not quite the same definition of professional institutions as the authors of the first directories. While these restricted their considerations to typical professional observatories where astronomers were located at that time, we took into account that nowadays many astronomers and/or space scientists can be found also in university departments or groups as well as in private companies. Sometimes also, it is difficult to draw a line between astronomy, space sciences and related activities.

22–5 A question of quality

The original philosophy of these directories was to provide practical data which one seeks always to have at one's disposal. They have proved over the years to be not only valuable auxiliaries for improving national and international relationships in professional and amateur astronomy, but also efficient tools for helping amateur astronomers, laymen and public bodies to contact astronomical organizations easily. Professional institutions have also stressed the utility of IDAAS for promoting astronomical popularization and for intensifying collaborations between their astronomers and amateurs.

These directories have reciprocally taken advantage of the experience gained with each successive edition, especially in the development of techniques for collecting, verifying and treating the data. ASpScROW 1991 will be the result of more than 13 years of data collection and technical refinement.

To compile a directory of real value is indeed quite a different venture to just reproducing and distributing, with comments of greater or lesser interest, data collected indiscriminately from all available sources. If professional file construction techniques are necessary, they cannot spare the extensive background, unrewarding and very careful work which is indispensable for the compilation of a valuable directory. The definition of a

Astronomical Directories

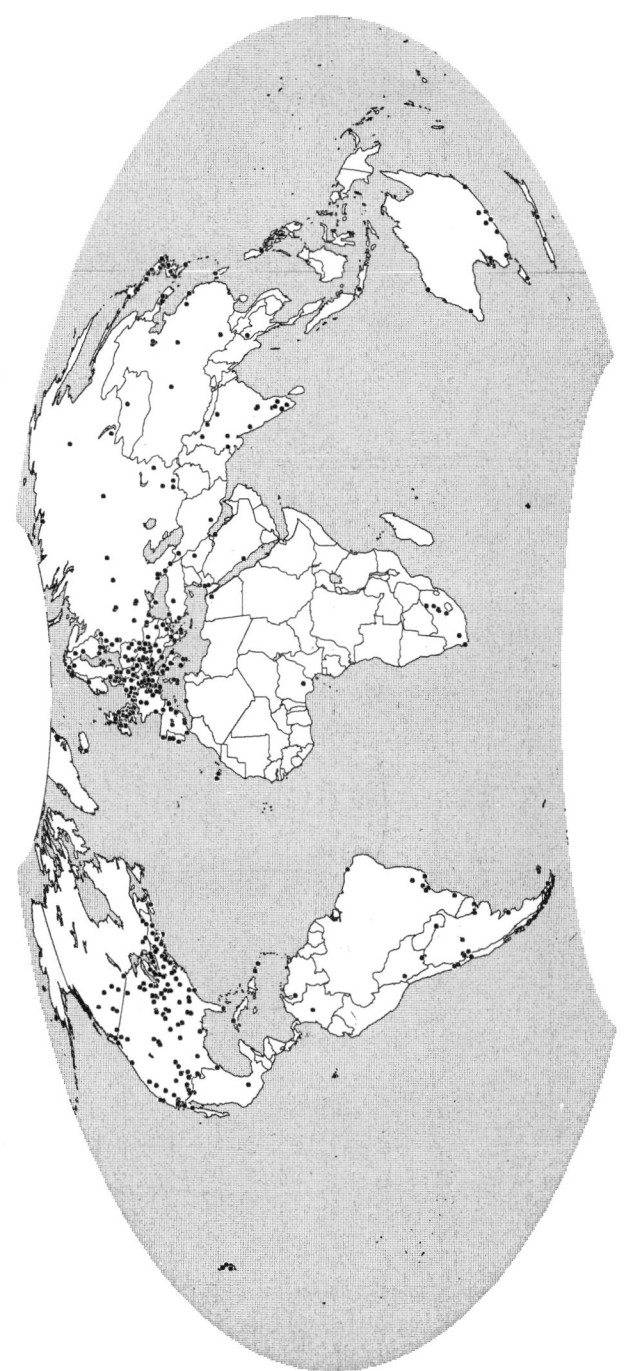

Figure 22-2: World distribution of astronomical and related institutions, including observing and receiving stations in ASpScROW 1991 (Heck, 1991) at the time of writing

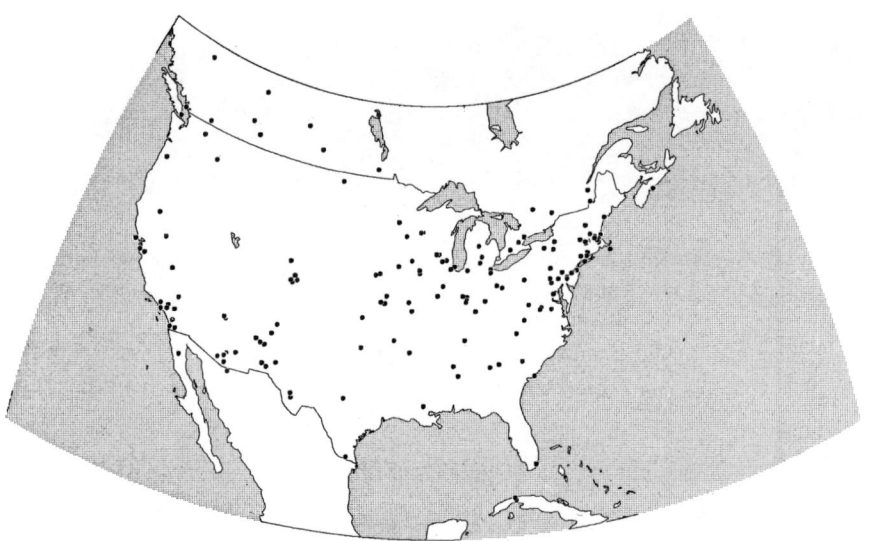

Figure 22–3: North American distribution of astronomical and related institutions, including observing and receiving stations in ASpScROW 1991 (Heck, 1991) at the time of writing

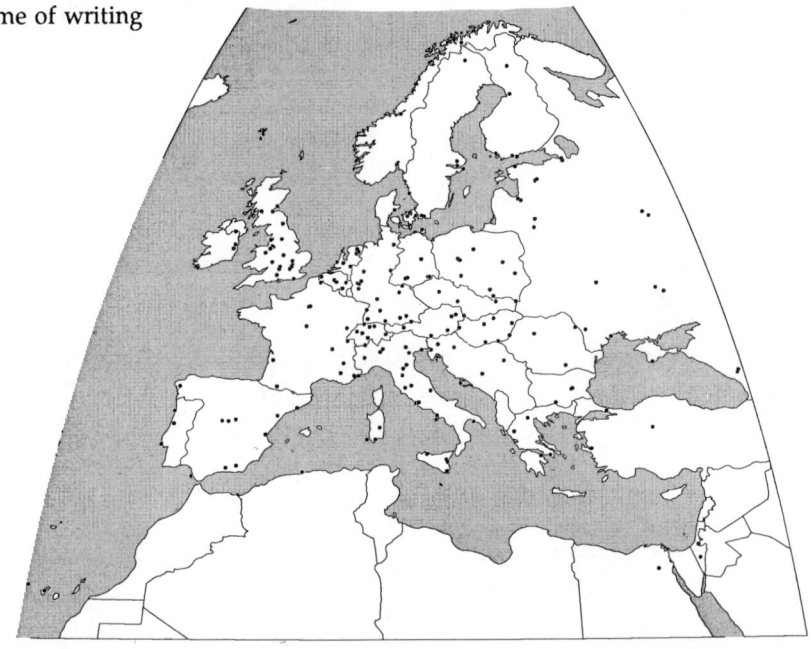

Figure 22–4: European distribution of astronomical and related institutions, including observing and receiving stations in ASpScROW 1991 (Heck, 1991) at the time of writing

very well profiled and adapted questionnaire, the homogenization of the data collected and the maximum reduction of the respondents' biases are all points that must be satisfied, often with the help of the most modern communication means such as electronic mail.

Moreover, it is imperative to take into account national differences: standards, conventions, habits, publication and financing channels usually vary from country to country. Finally, it is scarcely necessary to add that everything relating to professional astronomy can only be dealt with by professional astronomers, for evident reasons of competence and better knowledge of the realities of their corporation.

22–6 Speaking of evolution

The profile of the directories and the questionnaires sent to the various organizations listed have been improved and adapted along the years. The information gathered has been more and more comprehensive. As explained in the previous section, the categories of the entries listed in the directories have been also gradually broadened to better serve the needs of astronomers and space scientists.

One of the most striking changes in the course of the last years is the appearance of the electronic mail that has dramatically modified the way scientists communicate and exchange files of all kinds: a revolution in this era of communication similar to that of the computer earlier on. The way e–mail has appeared in IDPAI has been first as many addresses to different networks for only a few institutions. Then along the years, a few networks have emerged (as several of them also merged) and e–mail became widespread. Typically now, most institutions are connected, but to a maximum of two networks only on the average.

Telefax facilities have also become widespread recently. Telex and teletext seem consequently becoming less fashionable and disappearing slowly.

The political evolution in the world has also a direct impact on the directory contents, especially at the time of writing these lines while preparing also ASpScROW 1991. The liberalization of the political regimes in the European so-called Eastern countries has resulted in a dramatic increase of the questionnaires returned from the organizations concerned and an alteration of the list of the country names (as the disappearance of the German Democratic Republic).

Technical evolution have also been playing a significant role in the recent years and we went through them while producing the successive versions of our directories. Desktop publishing with all its flexibility and rapidity through the various packages available on mainframe, medium- and small-size computers did not exist when the directories mentioned in the first section of this chapter were produced: manual typesetting is dying, but it was the only way to get something printed a few decades ago.

Authors then had no access either to flexible database management packages as nowadays (do you still remember punched cards, not to speak of paper slips?). Think also how long could take a compilation of data through the postal service of that time when there was no jet transport, no satellite phone link, no photocopying machine, no... Really, Gentlemen, you deserve our admiration for what you did about a century ago!

Finally, let us stress that, if some trends are simply observed and recorded while others have to be anticipated, it is imperative to keep up with all of them in general.

22-7 Other directories

There are also quite a number of other directories in astronomy and space sciences and it would be difficult to quote all of them. The degree of detail in the information they provide is quite variable. Here are a few of the most significant of them.

Professional associations sometimes publish membership directories. This is the case for the *American Astronomical Society (AAS)* whose members are receiving each year a very useful directory where the AAS members are listed alphabetically, then grouped per institutions. A last section gathers the addresses of the North American institutions and of the most important ones from the rest of the world. In 1989, a complementary directory was published with electronic addresses. From 1990 onwards, these addresses have been included in the general directory (see AAS 1989 a&b, 1990).

The French *Société Française des Spécialistes d'Astronomie (SFSA)* has been issuing a directory, not only of its members, but of all astronomers and related scientists working in French institutions, as well as of French astronomers working abroad in institutions such as the European Southern Observatory (ESO) and the European Space Agency (ESA). Two editions have been published up to now (see for instance Egret, 1988) and a third one is in preparation. An important distinctive feature of these directories is to provide, beyond the usual individual data, a scientific profile for each entry and then an index according to the keywords listed.

The German *Astronomische Gesellschaft (AG)* has published once a membership directory (AG, 1978), but it seems that the experience has not been repeated since. The Dutch *Nederlandse Astronomenclub (NAC)* has also its directory (see for instance NAC, 1984).

Last, but not least, the *International Astronomical Union* (IAU, 1989) has published the directory of all its members (about 7000). Beyond the addresses and telecommunication data, one can find also the membership of individuals to the IAU commissions, giving in such a way a gross scientific profile.

More specific directories deserve also a mention here as the *International Astronautical Federation Directory of Member Societies* (Almár, 1984), the first latin–american astronomical census (Bohigas & Sarmiento, 1989) and a list of Czechoslovakian astronomical institutions (Hollan, 1989).

The blossoming of electronic mail in the recent years has led to numerous compilations of electronic addresses that people have been making available to each other over the networks. The situation in this field is however evolving quite rapidly and many of these lists have become quickly obsolete. Network management teams have been making available directories of their nodes and customers, both as files available on-line and on paper. See, for instance, Peters et al. (1989) for SPAN.

The most exhaustive compilation of e-mail addresses has been carried out by Benn & Martin (see for instance 1990) who are producing an *Electronic Mail Directory* essentially centered on individuals (about 5,000). Institutions (about 400) have also their communications 'plugs' listed and a long first section is devoted to the description of the most important networks and how to use them.

We wonder however whether such directories will not be absorbed progressively into more general ones (as it has been the case for the AAS). Their advantage is that they include not only members of specific societies, but also young researchers, associated engineers or technicians as well as computer scientists and managers.

On the amateur/grand public front, a number of compilations can be mentioned as those from the journal *Sky & Telescope (S&T)* (see for instance S&T, 1989), the *Federation of Astronomical Societies'* handbook (see for instance Jones, 1985), the *International Planetarium Society's* directories (see for instance Petersen, 1988), and so on.

Marx & Pfau's (1979) description of astronomical observatories is to be mentioned for German–reading aficionados. Kirby-Smith's directory and travel guide of US observatories (1976) is more tourist–oriented, while Krisciunas' (1988) *Astronomical Centers of the World* is definitely historical.

We cannot close this section without mentioning the excellent compilation on radio astronomy observatories carried out by the Committee on

Radio Frequencies of the US National Research Council (1989). Meszaros (1986) had also produced for NASA an atlas of large optical telescopes.

22–8 Future trends

Years ago, astronomers felt the necessity to set up computerized data banks and archives to cope with the accumulation and diversification of data and to get an easy and efficient access to the various types of data for a given object as well as to the corresponding bibliography. We shall not detail here the underlying philosophy (essentially centered on motivations for preliminary studies and on avoidance of duplications both of observations and analyses): please refer to the other chapters of this book that describe the various achievements and review also the status of the networking of the corresponding databases and archives.

In the present era of communication, astronomers, as other scientists, are now going a step further and break the geographical barriers through as many communication links as possible, a need that was already felt at the beginning of the century. Finalizing a joint paper by authors working at distant institutions does not need more than a few days now by exchanging the manuscripts, corrections, amendments and comments over the networks. Think what it was only a few years ago by normal postal service and over noisy phone lines and manned switchboards...

Consequently, keeping up to date directories with individual and institutional data, and especially those relative to the communication 'plugs', is more important than ever for the dynamism of our corporation, as well as for our investigations and projects. But it is not enough to keep these data up to date. They need to be made easily accessible to the community just by using these same communication facilities.

Therefore directories such as IDAAS, IDPAI, ASpScROW and membership lists such as the AAS and SFSA ones should be set up as on–line and interconnected databases. Other national societies should be encouraged to compile practical directories as complete as possible both in data and entries (not only members of the societies, but also young atronomers, engineers, technicians, and so on, when appropriate). Each of the organizations involved would be responsible for keeping its own directory up to date. Large institutions could also have their internal phone and/or e-mail directories plugged into such a service. This would be particularly useful for those using an individual direct–access system.

We heard with pleasure of the intent to set up a European professional Astronomical Society. In our opinion, one of its first tasks should be to set up a databank with all data available on the European astronomical and related personnel. By connecting such a network to scientific databases (through ESIS for instance where a *Users Directory* is already accessible on line), one can imagine that someone searching the bibliography in SIMBAD can immediately get the communication link data of one author he/she wants to get in touch with. The same could apply with someone detaining the proprietary rights on some images from a given archive.

Related activities can also be considered. Presently mailing stickers bearing the addresses of the organizations listed in the IDAAS, IDPAI and ASpScROW master files (permanently updated) have been made available, with a possible selection of countries and/or categories of entries. Other directories could offer the same facility for individuals and a selection on fields of interest should also be made available as an additional flexibility when appropriate. This would be particularly useful for meeting organizers, but also for publishers, manufacturers, dealers, and so on.

A directory such as ASpScROW provides also to the individual scientists the necessary data to contact easily and directly these publishers, manufacturers, dealers, software producers, etc.

References

[1] Almár I. (1984), International Astronautical Federation Directory of Member Societies, Hungarian Astronaut. Soc., Budapest, 44 p.
[2] American Astronomical Society (1989a), Membership Directory, Amer. Astron. Soc., Washington, 162 p.
[3] American Astronomical Society (1989b), Electronic Mail Directory, Amer. Astron. Soc., Washington, 62 p.
[4] American Astronomical Society (1990), Membership Directory, Amer. Astron. Soc., Washington, 232 p.
[5] Astronomische Gesellschaft (1978), Mitgliederverzeichnis, *Mitt. Astron. Gesellschaft,* **44**, 205.
[6] Benn Ch. & Martin R. (1990), Electronic Mail Guide and Directory, Roy. Greenwich Obs., Herstmonceux, e–mail files .
[7] Bohigas J. & Sarmiento A. (1989), Primer Censo Latinoamericano sobre Astronomía, floppy disk files.
[8] Committee on Radio Frequencies (1989), Radio astronomy Directories, Nat. Acad. Press, Washington, x + 80 p.
[9] Egret D. (1988), Annuaire de l'Astronomie Française 1988, Ed. Soc. Française Spécialistes Astron., Paris, 148 p.
[10] Heck A. (1989a), International Directory of Astronomical Associations and Societies together with related items of interest – IDAAS 1990, *Publ. Spéc. Centre Données Strasbourg,* **13**, vi + 716 p.
[11] Heck A. (1989b), International Directory of Professional Astronomical Institutions together with related items of interest – IDPAI 1990, *Publ. Spéc.*

Centre Données Strasbourg, **14**, vi + 658 p.

[12] Heck A. (1990), Acronyms and Abbreviations in Astronomy and Space Sciences, *Publ. Spéc. Centre Données Strasbourg*, **15**, ii + 400 p.

[13] Heck A. (1991), Astronomy, Space Sciences and Related Organizations of the World – ASpScROW 1991, *Publ. Spéc. Centre Données Strasbourg*, **16**, in preparation.

[14] Heck A. & Manfroid J. (1977), Atlas Photographique Astronomique, éd. Desoer, Liège, 224 p.

[15] Hollan J. (1989), The Astronomical Institutions in Czechoslovakia, floppy disk files.

[16] International Astronomical Union (1989), Alphabetical List of Members, 334 p.

[17] Jones B. (1985), Handbook for Astronomical Societies, Fed. Astron. Soc., Bradford, 68 p.

[18] Kirby–Smith H. T. (1976), U.S. Observatories: A Directory and Travel Guide, Van Nostrand Reinhold Co., New York, 174 p.

[19] Krisciunas K. (1988), Astronomical Centers of the World, Cambridge Univ. Press, Cambridge, x + 320 p.

[20] Marx S. & Pfau W. (1979), Sternwarten der Welt, Heider, Freiburg, 200 p.

[21] Meszaros S. P. (1986), World Atlas of Large Optical Telescopes, *NASA Technical Memorandum* **87775**, iv + 38 p.

[22] Nederlandse Astronomenclub (1984), Adresboek, 49 p.

[23] Pecker J.Cl. (1979), Report IAU Comm. 5 on Documentation, in *Reports on Astronomy – Vol. XVIIA – Part 1*, E. Müller (Ed.), D. Reidel Publ. Co., Dordrecht, 7.

[24] Peters D. J., Sisson P. L., Green J. L., Thomas, V. L. (1989), Space Physics Analysis Network Node Directory (The Yellow Pages), *NSSDC/WDC-A R&S*, **89–14**, 96 p.

[25] Petersen M. C. (1988), 1988 IPS Directory of Planetaria and Planetarians, Int. Planetarium Soc., Salt Lake City, 142 p.

[26] Rigaux F. (1959), Les Observatoires Astronomiques et les Astronomes, Obs. Roy. Belgique, Bruxelles, 460 p.

[27] Rigaux F. (1961), Les Observatoires Astronomiques et les Astronomes – Supplément, Obs. Roy. Belgique, Bruxelles, 42 p.

[28] Sky & Telescope (1989), Astronomy Resource Guide, Sky Publ. Corp., Belmont, 24 p.

[29] Stroobant P., Delvosal J., Delporte E., Moreau F., Vanderlinden H. L. (1931), Les Observatoires Astronomiques et les Astronomes, Casterman, Tournai, 316 p.

[30] Stroobant P., Delvosal J., Philippot H., Delporte E., Merlin E. (1907), Les Observatoires Astronomiques et les Astronomes, Hayez, Bruxelles, 318 p.

23

Science networks:
A short overview

Fionn Murtagh[1]
Space Telescope–European Coordination Facility
European Southern Observatory
Karl–Schwarzschild Straße 2
D-8046 Garching bei München, Fed. Rep. of Germany

23-1 Introduction

This chapter aims to give a user-oriented view of some current networking developments. A recommendable background article on networking, which is still very topical, is Quarterman & Hoskins (1986). We focus on the European situation, and point frequently to American developments.

23-2 Wide-area networks

The user should know about the network being used, and its attendant protocols, for the following reasons: (i) some networks have established greater reliability than others: for instance, some UNIX-based uucp links are dependent on telephone contact between computers which may not be operational at certain times; (ii) as a general rule, it is safer to keep the use of address routing through gateways between networks to a minimum: these ought not to be, but can be, a source of problems; (iii) store-and-forward type networks (EARN/Bitnet, for example) differ in the way the user sees them compared to a network which establishes a virtual link before sending a message (*e.g.* SPAN); and (iv) unfortunately it is still all-too-common for users to pass on their network addresses in a slightly garbled form (consider even JANET conventions for the order of parts of

---

[1] Affiliated to Space Science Dept., Astrophysics Div., European Space Agency.

the node name), and experience with networking helps to quickly resolve such problems.

Recent years have seen a consolidation of networking, rather than major new developments. Among principal wide-area networks (see Fig. 23–1) are the following:

- The Internet, which extensively covers the United States. It arose out of Department of Defense funding (through the Defense Applied Research Projects Agency (DARPA)), and went earlier under the name of ARPAnet. Today, the largest number of circuits are funded by the National Science Foundation (NSF) under the acronym NSFnet. Significant funding in recent years, involving especially the provision of high-speed trunk backbone links, owes its justification to the use of supercomputers accessible through the Internet. It is subdivided in domains, —educational (.EDU in the node name), military (.MIL), commercial (.COM), and governmental (.GOV). The protocols used by the Internet at ISO/OSI levels 3 and 4 (more of which below) are TCP/IP (Transfer Control Protocol/Interaction Protocol). The NASA Science Network (NSN) belongs to the general Internet.
  Among services offered by the Internet are remote logon (telnet), file transfer (ftp), and restricted access logon (anonymous ftp). TCP/IP protocols are also used by a small but non-negligible number of European networks, including NORDUNET in Scandinavia, SWITCH in Switzerland, and others (Fluckiger, 1988, p. 29).
- BITNET ("Because It's Time Network": Fuchs, 1983), which goes under the name EARN (European Academic and Research Network) in Europe, and Netnorth in Canada, originally relied upon leased lines made available by IBM. In Europe, the protocol used was Remote Spooling Communication Subsystem (RSCS) the design of which predated modern layered models. EARN/BITNET is a store-and-forward network which only allows file transfer and mail. Emulation of RSCS on non-IBM machines is carried out, for instance, by software packages on Vax/VMS machines such as NJE (public domain, from Argonne National Laboratories) or JNET (from Joiner Associates, Inc.).
- Space Physics Analysis Network (SPAN; Green, 1988) is a network based on DECnet protocols, and thus mainly used between DEC machines. SPAN in the U.S.is funded by NASA and headquartered in Goddard Space Flight Center. In Europe, the SPAN *backbone* network is largely funded by ESA.DECnet can be run on dedicated lines or over public switched networks. At application level, it operates on the principle of a link being established, the message file sent, and finally the link closed. SPAN is a popular network among astronomical institutes with more than 1000 systems interconnected. A set of back-

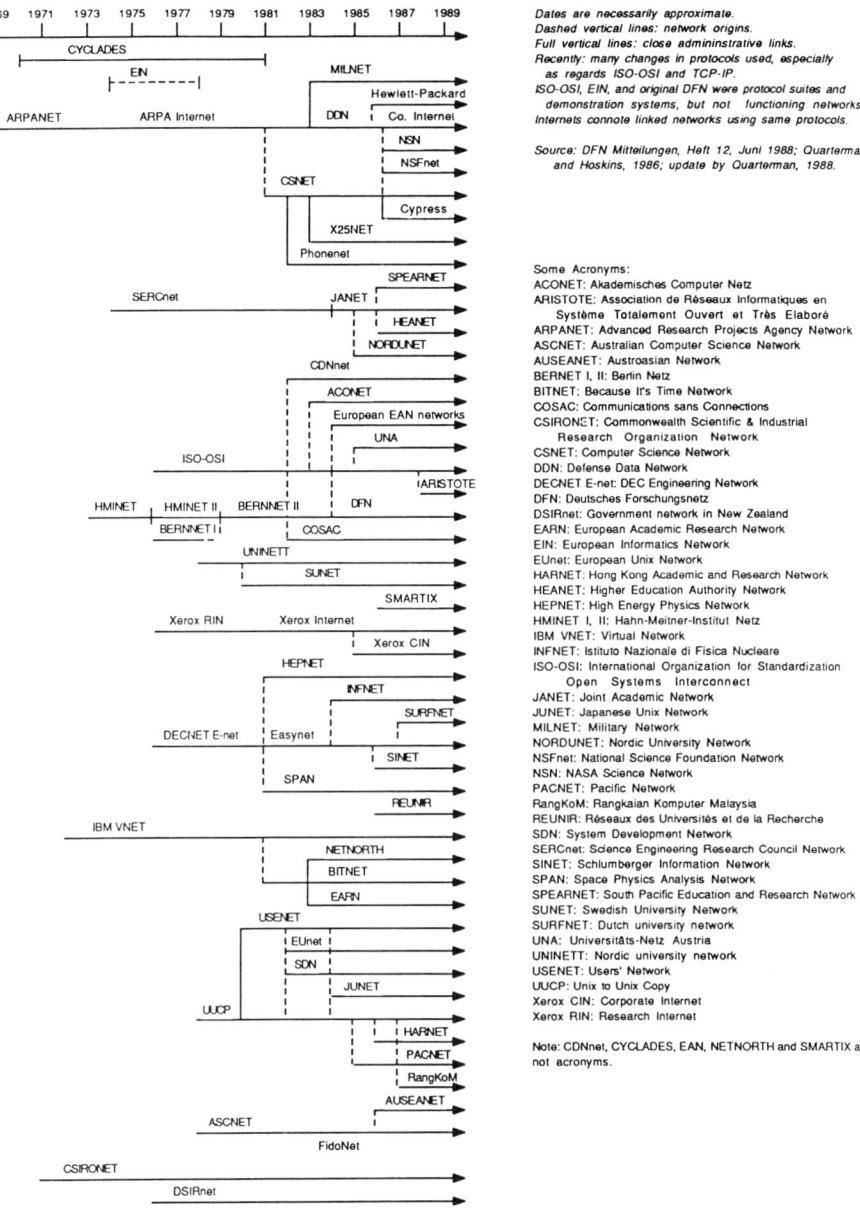

Figure 23–1: Notable networks.

bone links between routers at Goddard Space Flight Center, Marshall Space Flight Center, Johnson Space Center, and the Jet Propulsion Laboratory provide 56 Kbits/s transfer. Tail circuits connect a given institute to a router or to the backbone, generally using leased lines at 9.6 Kbits/s. The European SPAN (E-SPAN) backbone has reached 32 Kbits/s bandwidth between ESTEC, ESOC, ESA Headquarters and ESRIN, the four major ESA establishments.

- National networks: using very different protocols, such networks are often the primary beneficiaries of funding in Europe. JANET in Britain, based on the Coloured Books protocols, early on established broad coverage of universities and other institutes. It uses domain names (.AC.UK, for academic community; .MOD.UK, for defence ministry, .CO.UK, for commercial), the parts of which, a little confusingly, are usually referred to in the reverse order internally among JANET users. Other countries in Western Europe have set up networks with node addresses which include a reference to the country: .FR in France, .DE in Germany, .SE in Sweden, .PT in Portugal, .IE in Ireland, .IT in Italy, *etc.* The protocols differ, but gatewaying and routing are carried out in a manner transparent to the user. Recently, a national SPAN sub-network has been setup linking a considerable number of astronomical centres plus the Centre National d'Etudes Spatiales (CNES). Based on such networks, a number of discipline-oriented networks are available. Among the most notable we mwntion here the astronomical STARLINK and ASTRONET networks, in Britain and Italy, respectively. Both of these encompass activities additional to computer networking – software coordination, workshop organization, and so on (see chapters 18 and 19, this book, for an overview on their database activities).
- Thematic networks: the NASA Science Internet (NSI) embraces SPAN and the NASA Science Network (NSN), a TCP/IP network. NSI is used by over 10000 people for NASA-related projects alone (Rounds, 1989). The DECnet-based High Energy Physics network (HEPnet), and MFEnet (magnetic fusion energy) are also heavily used in their respective communities. CSnet, originally a mainly U.S. computer science network using various protocols, has merged with Bitnet.

23-3 Organisations

Communication between equals requires standards —the rules of debate, so to speak— and computer communications strongly embody this principle. The National Science Foundation is currently the leading player in the United States. A nationwide backbone offering "T1" speed (*i.e.* 1.544

Mbit/s transfer rate) is expected to be in operation within two years. Within five years 45 Mbit/s links are expected, and by the end of the century 1–3 Gbit/s (based on fibre optic connections). If successful in obtaining budgetary appropriations in 1991, the proposed National Research and Education Network (NREN) will replace the NSFnet, MILnet, ARPAnet and other Internet (TCP/IP protocol) networks, and proceed with these developments.

Country	Organisation	Coutry	Organisation
Austria	ACONET	Netherlands	ERNET
	UNA		HBONET
Denmark	DENET		PICA
	DUNET		SURF
Finland	FUNET	Norway	BIBSYS
France	ARISTOTE		UNINETT
	REUNIR	Portugal	RIUP
Germany	DFN	Spain	FAENET
Greece	ARIADNE		IRIS
Iceland	SURIS	Sweden	OSINET
Ireland	HEANET		SUNET
Italy	CILEA	Switzerland	CHADNET
	CINECA		SWITCH
	CNR	UK	JANET
	CSATA	International	CERN
	ENEA		EARN
	INFNET		EUNET
			NORDUNET
			E-SPAN

Table 23–1: Networking organisations and communities in Europe. (Based on Bauerfeld, 1988, p. 6.)

Among networking standards, the International Organisation for Standardization (ISO) in Geneva plays the particularly important role of coordinating the Open Systems Interconnect (ISO/OSI) standards. The ISO seven-layer model of what is involved in sending a message from one partner to another considers the tasks as ordered in seven layers: physical, data link, network, transport, session, presentation, and application. The X.25 series of standards relate to the lower three levels. This X.25 infrastructure is now commonly available. It is used directly by the user whenever he/she makes use of public packet switching networks (for remote logon, for instance), or increasingly often as the substratum for a wide-area network.

In Europe, many organisations are active in furthering networking (Table 23-1). JANET in Britain, Réunir in France, Deutsches Forschungsnetz (DFN) in Germany, and the Gruppo Armonizzazione delle Reti per la Ricerca (GARR) in Italy, are some of these. The DFN in Germany support the Wissenschaftsnetz (WIN) which offers better tariff policy compared to the public X.25 network. WIN, which is currently being installed, offers 64 Kbit/s and was expected to upgrade to 2 Mbit/s during 1990. Using Wissenschaftsnetz, the German part of EARN will migrate towards ISO/OSI protocols. This means first and foremostly the use of X.25. The IXI backbone, discussed below, will be of major benefit here. Further plans are for X.400 (electronic mail) and other OSI-conform products (Anon., 1990).

At European level, Réseaux Associés pour la Recherche Européenne (RARE) coordinates practical networking activities at a general level. Membership is on a national basis, and additionally includes international organisations (*e.g.* Centre Européen de Recherche Nucléaire (CERN); European Centre for Medium-Range Weather Forecasting (ECMWF); European Space Agency (ESA)). Major projects of RARE, currently in progress, include a message handling system (MHS) pilot project, testing X.400 electronic mail standards and implementations; the Eureka project COSINE (Cooperation for Open Systems Interconnection Networking in Europe), to facilitate vendor-independent networking; and the IXI project (International X.25 Interconnect). The latter aims at providing a universal 64 Kbit/s (and later 2 Mbit/s) X.25 backbone in Europe.

23-4 The future

Improvements in speeds of data transfer are taking place continuously. However while many international links allow 64 Kbit/s (as against 9.6 Kbit/s a few years ago), this is still a long way from the 10 Mbit/s to which the local area networker is used to (Brinkhuijsen, 1988, p. 5).

In Europe, the coming period will undoubtedly see increasingly better connectivity to eastern Europe and the Soviet Union. The coming years will see astronomical work being channelled more towards telescience and work with remote databases (see elsewhere in this volume). The very useful hard-copy directory produced by Benn and Martin (1990) will be complemented by more "white pages" on-line equivalents (see also chapter 22, this book).

Many users still use telephone lines (*e.g.* using a modem or acoustic coupler) or, directly, an X.25 packet-switched connection (*e.g.* DATEX-P in Germany, Transpac in France, Itapac in Italy, *etc.*). As the telecom companies in Europe slowly begin to turn from being industrial concerns to being

Country	Researchers	Students
Austria	6,710	165,310
Belgium	13,880	245,760
Denmark	10,000	113,160
Finland	10,950	119,520
France	72,890	1,144,080
Germany	140,000	1,503,040
Greece	2,630	124,700
Iceland	300	5,130
Ireland	2,770	64,120
Italy	63,020	1,181,950
Netherlands	21,550	384,130
Norway	7,750	91,330
Portugal	3,020	99,170
Spain	14,230	787,860
Sweden	17,040	221,000
Switzerland	16,410	105,900
Turkey	13,880	Unknown
UK	86,500	897,000
Yugoslavia	24,880	375,390

Table 23–2: Networking needs in Europe: researchers and students enrolled in the COSINE (Cooperation for Open Systems Interconnection Networking in Europe) states. Potential networkers: 530,000 researchers, 530,000 research support staff, and 50% of students, *i.e.* 3,500,000; in total 4,500,000. Source: Bauerfeld, 1988, pp. 10–11.

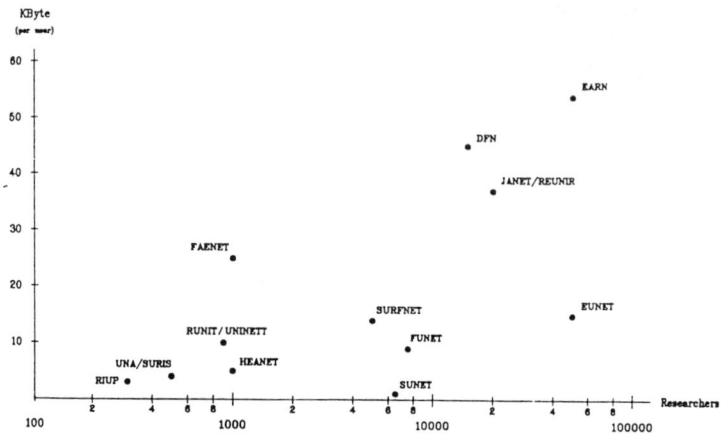

Figure 23-2: Number of users versus average amount of traffic (Kbytes) per user. Source: Bauerfeld, 1988, p. 15.

service-oriented, there will be an explosive evolution in consumer possibilities. The digitalization of Europe's telephone network is expected by 1993. Numeris in France, and Integrated Services Digital Network (ISDN) in Germany and elsewhere, promise data transfer at 64 Kbit/s, and the integration into one network of services which currently rely on digital and analog circuits. Apart from many new telephone services, the new S_0 interface (upgrading the widely-used V24 or RS232 interface) will allow faster data, speech, text, and image transfer, and the subsequent facilitating of the convergence of the areas represented by these.

The transfer of information, as every astronomer knows, does not stop with voice, text and data. Image transfer is currently not easy: physical transport using magnetic or optical devices is still the order of the day. The phenomenal growth of facsimile (fax or telefax) communication indicates the view of the consumer-in-the-street.

The growth of fax communication is mainly based on Group III machines. These provide up to 9600 baud transfer, implying up to four pages/minute. Group IV machines go considerably beyond this, offering 64 Kbit/s, *i.e.* three seconds per page, and using plain paper (Law, 1989). The latter standard is optimized for digital public switched telephone networks —for example, ISDN. It is foreseen that Group IV machines, with ISDN, will bring us considerably towards the goal of having a single piece of equipment perform photocopying, fax, telex, videotex, email, laser printing, text formatting, picture handling, *etc.*

Acknowledgements

For discussion and references, I am grateful to R. Hook and F. Pasian. Useful suggestions from the editors were highly appreciated.

References

[1] Anon. (1990), The German network for science, *IES News*, **27**, April 1990, 1–2.
[2] Bauerfeld W. (1988), *COSINE Specification Phase. Report 3. The Scope of COSINE*, in RARE (Ed.), *COSINE: Cooperation for Open Systems Interconnection Networking in Europe. Eureka Project 8. Specification Phase*, Commission of the European Communities.
[3] Benn C. & Martin R. (1990), *Electronic Mail: Guide and Directory*, 1 Jan. 1990 version, Royal Greenwich Observatory.
[4] Brinkhuijsen R. & Fluckiger F. (1988), *COSINE Specification Phase. Report 6. Public Services*, in RARE (Ed.), *COSINE: Cooperation for Open Systems Interconnection Networking in Europe. Eureka Project 8. Specification Phase*, Commission of the European Communities.
[5] *Computer Magazin* (1990), **3/4**, April 1990 *Die Zukunft der Telekommunikation: ISDN*, pp. 1–2; *ISDN-Gebühren unter der Lupe*, O.F. Schröter, pp. 24–27.
[6] Fluckiger F. (1988), *COSINE Specification Phase. Report 5. Future Services in COSINE*, in RARE (Ed.), *COSINE: Cooperation for Open Systems Interconnection Networking in Europe. Eureka Project 8. Specification Phase*, Commission of the European Communities.
[7] Fuchs I. (1983), *Bitnet*, **Perspectives in Computing**, 3, 16–27.
[8] Green J. L. (1988), *The Space Physics Analysis Network*, **Computer Physics Communications**, **49**, 205–213.
[9] Huton J. (1989), *Future scientific networking*, **Computer Physics Communications**, **57**, 188–190.
[10] Law C .E. (1989), *X.400 and OSI. Electronic Messaging into the 1990s*, IBC Technical Services Ltd., 57-61 Mortimer St., London W1N 7TD.
[11] Murtagh F. (1986), *A guide to computer communications and data networks for astronomers*, Bull. Inf. CDS, **31**, 89–119, 1986.
[12] Murtagh F. (1988), *Data and networks*, in C. Jaschek and C. Sterken (Eds.), *Coordination of Observational Projects in Astronomy*, Cambridge University Press, Cambridge (UK).
[13] Quarterman J. S. & Hoskins J. C. (1986), *Notable computer networks*, Communications of the ACM, **29**, 932–971.
[14] Rounds F.N. (1989), *NSI's management perspective on its evolving networks*, Information Systems Newsletter, Nov. 1989, 43–35.

24

User Interfaces in Astronomy

*Fabio Pasian & **Alan Richmond[1]

*Space Telescope–European Coordinating Facility
c/o European Southern Observatory (ESO)
Karl–Schwarzschild–Straße 2
D–8046, Garching bei München
Fed. Rep. of Germany

**European Synchrotron Radiation Facility
BP 220
F–38043 Grenoble Cedex
France

24-1 Introduction

The user interface is a fundamental component of information systems: the extent to which information system functions are useful is critically measured on how easy they are to use for both novices and experts. However, it is usually very difficult to define in a clear way what a "good" user interface really is, because this definition is often a matter of personal taste; therefore, the ideal interface should be able to change in aspect and response, in order to cope with rapidly changing user needs.

User interaction with computer–based systems in the astronomical domain is in general different from system to system. Even in the database and archive area, where it could be thought that a certain degree of commonality exists, there is a wide range of different user interfaces, often hampering the work of one-time-users. While in the database field the lack of homogeneity has been recognized as a problem and is being tackled by information systems projects such as ESIS and ADS (see chapters 14 and 15 respectively, this volume), in other fields the issue is still wide

---
[1]Part of this author's contribution was written while he was at the Space Telescope Science Institute, Baltimore, MD, USA.

open.

In general, the software required to manage the user interface contributes largely to system complexity and development cost, both in manpower and financial terms. In the design and implementation of a system, the interfaces between software modules usually are clearly defined at an early stage. On the contrary, the interface between the computer and the user may not be clear until shortly before installation; even afterwards, when functions are tested in a "real–life" interaction with the system, interface design needs to be continuously refined. When planning and designing a software system for any astronomical application, this consideration has to be kept in mind, so as to avoid unpleasant surprises during the implementation phase.

In this chapter, an overview of different user interfaces concepts is given, ranging from the simplest *prompt–and–answer* teletype mode to the most sophisticated window-based modes, allowing integration of graphics and image display. Examples related to specific systems existing and available to the astronomical community are given. Finally, a discussion on modern software technologies for the construction of user interfaces is also included. This section is rather technical and may be of little interest to the *user* of an astronomical software system; *designers* and *developers* will probably find it interesting, since up-to-date concepts, techniques and tools for user interface design are discussed in detail.

24–2 Current Status

2.1 Teletype-mode interfaces

Most of the user interfaces astronomers are used to dealing with, are the simplest, running on alphanumeric terminals in a teletype mode, following the *prompt–and–answer* or *command–mode* paradigms. There are some historical reasons for this choice, since up to seven-ten years ago the usage of "dumb" terminals (teletype-mode, no alphanumeric cursor capabilities) was very common, and in general no sophisticated or standard graphical tool was available.

Some of these considerations still hold true in the case of systems being accessed using wide-area network connections. Availability of wide-band connections is not very common yet, and in this case sophisticated user interfaces may lead to unacceptable response times, since screen "painting", or transmission of graphic protocols usually result in very inefficient usage of the network.

The basic mechanism behind a *prompt–and–answer* interface is extremely simple: questions are asked and the user provides answers following pre-defined rules. In other cases, the user needs to input a full command line in a specified syntax. Defaults values for the parameters to be provided are sometimes available, and help, if existing as a feature, can be obtained by typing a special character, such as H or ?. The result is far from being fancy, and the amount of user-friendliness can be fairly limited, but this concept has the advantage of allowing the interface to be run on *any* kind of terminal.

The software architecture used to implement a teletype-mode interface is rather simple: the prompting and answer-capturing mechanism is implemented in the code by means of simple write-to-terminal and read-from-terminal routine calls, handling special cases such as defaults and help.

There are many examples of simple *prompt–and–answer* interfaces in the astronomical environment, especially in systems having been designed and implemented during the early 80's or even before. A system which is widely known and used for remote user access is the interface to the Uniform Low Dispersion Archive (ULDA) of the IUE observations (Wamsteker 1989), which has been installed on a number of national hosts throughout Europe and in some non-European countries (see also chapter5 in this volume).

The access to ULDA is made through USSP (ULDA Support Software Package), which is basically composed of two parts, one running at the national hosts, and one running on the remote user's own computer. The QUEST program, running at the national hosts, can be considered as the user interface to the system (Driessen & Pasian 1988). After validation of the user's identification (for statistical purposes only), the interface offers a choice of possibilities to select from the whole archive the data needed and to prepare data as a compressed file for downlink. To select data, a number of questions such as object coordinates, IUE camera image and aperture parameters, object class and homogeneous object id are asked. The user may answer by typing in valid values, skip the answer by pressing the <Return> key, or ask for help by hitting the ? key.

There is no sophistication whatsoever in the user interface, and there is a strong rationale behind this approach. The ULDA national hosts are remotely accessed using any type of network connection possible: dedicated lines, commercial packet switching, dial–in modems. It is obviously clear that, while some connections are very efficient, some other can be extremely slow and unreliable. For these reasons, the user interface to the system has been kept simple, and there is even an option to obtain short prompts for experienced users to minimize the screen-filling time.

2.2 Interfaces with graphical output

The basic extension to the simple prompt–and–answer interfaces is the possibility to have, on pre-specified computer peripherals or on workstations, some graphical output allowing the user to have a better understanding of the system or of the data being dealt with. As an example, it is in this way possible, after extracting a table from a database, to plot data contained in one column versus data in one of the other columns.

The hardware requirements to be able to use such an interface exceed of course the simple teletype-mode terminal: the absolute minimum required is a terminal with graphic capabilities, compatible with the graphic protocol used by the system. The software architecture used to implement a command-mode interface with graphical output is not much more complicated than the one for the prompt–and–answer mode. As a matter of fact, additional modules are needed to provide the output in graphical format: in principle, the graphic modules will address a device-independent graphic library which will then take care of dialogue with the terminals.

This approach has been followed in building the user interface for the remotely-accessible EXOSAT database (see chapter 2, this volume). The EXOSAT user interface is a command-driven environment allowing to "browse" through the data and to select the required ones. Graphic capabilities are provided to allow the user to display the data retrieved, in order to select those to be possibly downlinked to his/her own computer. An additional possibility is to redirect the graphic output to produce Postscript files, which can also be sent to the user, giving publication-quality graphs.

To optimize transmission times over the network, fairly cryptic commands are used; but the graphic system has the capability of minimizing the graphic sequences sent to the terminal, and this feature allows display of graphs with reasonable response times even via wide area networks.

2.3 Interfaces allowing interaction with data

Once the user has access to data on his/her own computer, he/she may want to analyse them and/or interact with them, *e.g.* choosing specific subareas of an image, or inquiring for the value of a specific pixel, or choosing a specific set of spectral lines to be subsequently fitted or compared with an existing model.

Such facilities are available on all of the currently used data analysis systems, such as MIDAS, IHAP, IDL, IRAF, STSDAS, AIPS, *etc.* The way to specify the command to be executed is still of the *command–mode* type, but image display and/or graphic capabilities are available, and interaction with the data is allowed by means of cursors. This feature allows the user to give input to the system in an alternate way, rather than just via a terminal and a keyboard.

Of course, the level of the hardware needed to run user interfaces allowing interaction with data is much higher than in the preceding cases. This kind of user interface may be implemented either on a set of different peripherals, such as graphic terminals, image display devices and the like, or on the bitmapped screen of a workstation, where a number of virtual devices (alphanumeric, graphic, image display) are opened to support operations.

The software architecture of such systems gets rather complex, since different libraries are required to handle alphanumeric dialog and command input, graphic functions (output on screen, input from different sources—cursor, keypad, keyboard, mouse buttons) and, similarly, image display functions. Usually, in order to neatly separate the system components (alphanumeric, graphic and image display), each one is implemented as a different subprocess of the user parent process. In general, for portability concerns, device-independence and operating-system-independence is desired: the software then has a complex structure, where application, services and specific system dependent code are isolated in different layers.

The above-mentioned data analysis systems all share to some degree the same philosophy: MIDAS [14] will be used as an example for further illustration. MIDAS is a command-driven environment: therefore all of MIDAS facilities can be called by entering a command in a predefined format as an answer to the system prompt (Midas>). Specific commands will give access to graphic or image display facilities.

A typical graphic session on MIDAS could look like this:

> Midas> ASSIGN/PLOT *device-name*

assigns *device-name* to be the graphic device where all graphic I/O is to take place. Then,

> Midas> PLOT/CONTOUR *image-name*

plots the contour map of *image-name*, a two-dimensional image.

> Midas> GET/GCURSOR *table-name*

allows the user to use the arrow keys on a graphic terminal (or use the mouse on a workstation window) to move the cursor and be able to read back its coordinates in a MIDAS table called *table-name*. Or

> Midas> CENTER/MOMENT CURSOR *table-name*

allows him/her to use the cursor to visually identify the coordinates of central positions of objects; the precise positions are then computed using an intensity weighted first moment of the pixel values and are written in a MIDAS table called *table-name*.

Similar commands are available for the image display facilities. As an example, once an image (*image-name-1*) is loaded on the available image display device via the

```
Midas> LOAD/IMAGE image-name-1
```

command, the cursor on the screen can be defined as

```
Midas> SET/CURSOR cursor-number cursor-shape
```

or a subset of the currently displayed image can be extracted using the

```
Midas> EXTRACT/CURSOR image-name-2
```

command, which allows usage of the cursor(s) to define interactively on the screen the diagonal of the subimage to be stored in *image-name-2*.

2.4 Form–filling interfaces

All of the above user interfaces, although allowing rather complex graphical and image display output and interaction, are still based on the concept of using the teletype-mode prompt–and–answer system to accept user commands. This is still very popular in scientific environments, where application software is seldom kept up-to-date with the pace of technical development. In the computer market, user interfaces based on forms have massively appeared in the 80's and are most commonly available on PCs.

The concept underlying a *form–filling* interface is extremely simple and is based on the idea of filling a questionnaire: displayed on the screen in front of him/her, the user has a form having a number of fields. Such fields must or can be filled, *e.g.* to specify the parameters of a command, or to qualify a query on a database. This concept is particularly useful for the once-in-a-while user, because it relieves him/her from having to learn command syntaxes or having to browse through manuals.

Form–filling software can be very conveniently run on bitmapped workstations *e.g.* under the X-Windows standard; the existence of X-Windows toolkits furthermore allow such an interface to be very naturally implemented. However, the hardware requirements needed to run a form–filling user interface can be fairly low: a terminal with the capability of addressing an alphanumeric cursor on the screen and overwriting text is all that is needed. These capabilities are ANSI-standard: additional features, such as reverse, boldface and blinking, although pleasant, are non-required options. Since many years, most terminals (*e.g.* VT100 and clones) have all of these features: adaptability to specific terminals can anyhow be achieved using "termcap" files, containing the capabilities of a large number of terminals and allowing the system to adapt to the features of the specific terminal being used. Emulation of an ANSI-standard terminal is available also on workstation windows.

There are a number of astronomical applications using the concept of form–filling user interface. Most of those known to the authors (STARCAT, the DMF Operator Interface, SIMBAD 3, the PEPSI proposal entry assistant, the prototype ESIS User Shell) are based on the Proteus [21] and TermWindows [17] software libraries. As an example of the various applications, only the first two will be briefly described, since their scope and purpose are quite different.

STARCAT (Pirenne *et al.*, 1988 and [24,26]) is the user interface to both the ESO archive and the Data Management Facility for the Hubble Space Telescope (refer also to chapters 11 and 6, this volume), and allows access to about 40 astronomical catalogues. STARCAT is basically an interface connected with a database, where catalogues (*e.g.* standard astronomical ones, or the catalogue of HST observations) are stored. The goal of its design has always been user-friendliness; the user chooses from a menu, via the arrow keys on the terminal keyboard, the operations to be performed. To access the database management system, the user does not need to issue SQL statements but rather, after having chosen the qualify option in the menu, navigates in a form, filling the fields he/she want to constrain in the database search: the software interprets the contents of the form, generates SQL statements and sends them to the database management system. "Hot keys" are available at any time to perform or abort the specific action requested via the menu, or to obtain help, which is generously provided in the system.

The DMF Operator Interface (OIP – Operator Interface Program) (Pasian & Romelfanger, 1989 and Pasian & Vuerli, 1990) has been designed and implemented to supply the HST archive operators at STScI and ST–ECF with an easily-usable and reliable tool. The concept behind OIP is to reduce the probability of operator errors by means of form-based operator commands preparation. Each command can be selected on a menu via the arrow keys of the operator's keyboard: at this stage, a form to be filled appears on the screen. After the form is filled, OIP takes care of formatting the request to be sent to the rest of the DMF, checking the syntax prior to execution. OIP's concept is similar to STARCAT's since the tools used for its implementation (Proteus and TermWindows) are the same.

2.5 Highly-graphical interactive interfaces

The appearance on the computer market of bitmapped-screen workstations all sharing the same screen management software (X-Windows), and of low-cost but highly effective and easy-to-use systems, such as the Apple Macintosh, has led also in the astronomical environment to the development of new user interfaces based on these tools. Users will soon become familiar with concepts such as hyper-text, which allows the possibility of interactively (*e.g.* with a mouse) choosing a highlighted character string in a portion of text being displayed, in order to receive additional information, in textual, graphical or image form.

As examples of highly graphical and evolved user interfaces, reference to the PICKLES program running on an Apple Macintosh is made, followed by a brief description of the ESO remote observing user interface, currently run on a Ramtek display and being updated to X-Windows. Finally, reference to SAOIMAGE, an image display package taking full advantage of X-Windows capabilities, will be made.

PICKLES (McCartney 1990) is a program to determine pointing and rolls of HST when the target and guide stars are in the Fine Guidance Sensors or in the aperture of the HST instruments, extracting and displaying star positions from the Guide Star Catalog CD-ROMs. Pickles is a standard Macintosh application, written using an object-oriented approach: it makes full use of pull-down and pop-up menus, key commands and mouse actions and allows of course full graphic data manipulation.

The ESO remote control system is used at ESO headquarters in Garching to query and modify the status of some of the ESO telescopes at La Silla, Chile, through a 64 Kbit/s satellite link. In particular, the user interface for the NTT telescope is implemented on a Ramtek display and is currently run locally at the telescope site. The screen is divided in a number of areas (menu, status display, areas containing parameters to be monitored and/or to be modified, message area, alarms area). The user can select the various screens and actions by moving the cursor on the Ramtek display, can modify instrument or telescope status by updating parameters in the displayed forms, can monitor the status of the system by inspecting non-modifiable parameter tables or through the help of graphic facilities (*e.g.* checking the light path inside an instrument as a function of mirror tilting). Informational system messages and alarms (messages requiring immediate user action) are displayed in the message and alarm areas, respectively. An update of this user interface is currently being implemented in order to exploit the satellite link for remote operations. A new user interface following the same philosophy is being designed and will be implemented in 1991; it will be based on the X-Windows standard and will run on a Unix workstation.

SAOIMAGE is an application program written at the Smithsonian Astrophysical Observatory (SAO) as a tool to visually inspect images written in the standard FITS format. It has a very advanced and interesting user interface, since it exploits the capabilities of X-Windows: the mouse and the graphic cursor on the screen are used not only to allow full graphic and image display interaction, but also to choose actions to be performed, new options, new menus. As an example, the screen look-up table can be modified in real time by modifying by mouse action the red, green and blue contributions to the 256 levels of the colour look-up table. The system is definitely user-friendly and self-explanatory, besides being powerful for image inspection tasks.

24-3 Designing the User Interface Software

In the preceding sections, an attempt was made to give an overview, even if incomplete and possibly biased, of a number of different approaches to user interfaces, already available in the astronomical environment. But what if we were in the position of planning a new system, such as a new astronomical archive, or a facility allowing control, quick look and data preprocessing for a new instrument?

When planning a new software system for scientific users, the acquisition of suitable tools to perform flexible design and rapid prototyping of the user interface should be considered as a high priority topic, to allow an efficient design/implementation iterative phase. Underestimating the importance of such tools, or devoting a low level of resources to the user interface section of a project would be a mistake; after all, it is usability which gives a measure of the success of a system: and the user interface is the key to fully exploiting the capabilities of the system itself.

3.1 User Interface Management Systems

A major step in the direction of decoupling user interface programming from applications code, is to provide tools for the programmer to specify screen formats, menu and form layouts, prompt input validations, *etc.*, independently from the code; the tools *automatically* generate the user interface. Software supporting these capabilities has become known as a User Interface Management System (UIMS), by analogy with DBMS.

The separation of application tasks from user interface details relieves application programmers from programming the communications with the user, so that the applications code is very much simplified, and the development time is shortened (Bass, 1985, and Bass & Bunker, 1981). It also permits them both to evolve more independently, reduces coding, facilitates rapid prototyping, enhances portability and customisability, and encourages the development of reusable software (Richmond 1986).

UIMS's typically comprise an interactive graphics–oriented editor; a code generator (*e.g.* producing C); and a simulation facility so the interface can be incrementally prototyped. Some example UIMS's are:

tae+ : NASA's Transportable Application Environment (TAE+) contains a *Workbench* for prototyping and building graphical user interfaces (Götz, 1990); it is used to create and edit *panels*, which are X windows that can contain user interaction objects (*items*) such as buttons, icons, text, stripcharts, *etc*. Items can generate or respond to events; actions can be specified for events. Panels are described by *resource files* which can be transferred between computers in binary or ASCII format. Some of the other strong points of this product are that it is available on a wide range of platforms; is inexpensive; includes dynamic behaviour coding facilities; offers a solution for 2D synoptics; has a strong user community (esp. NASA); and will be Motif compliant in the 5.0 version promised for spring–summer 1991 (see section on toolkits below).

TeleUSE is Telelogic's User Interface Software Environment. It comprises an interface layout editor; a manager to control the dialogue between user and computer; an application interface manager to specify communication between dialogue control and application code; a runtime library; and a tool for compiling a description of the user interface.

3.2 Support Requirements

A comprehensive user interface supports all aspects of the interface, including screen display, and the user–application dialogue. Modern workstations facilitate this with bit mapped graphics, keyboard, and

Multitasking for multiple applications running concurrently;

Device independence allowing applications code to use a new hardware display unaltered;

Pointer devices *e.g.* mouse, so that menus, dialogue boxes, clickable active regions, *etc.*, can be used to eliminate the need for users to learn special command languages and obscure syntax;

Network transparency to permit access from the user interface over a network without requiring that a local copy of the entire system be available;

Windows organised in a hierarchy; overlapping, resizable; output to partially obscured windows; *etc*.

Screen Management *e.g.* for

- feedback to show that input has been received.

- display update when application data changes.
- field scrolling and editing.

Input Management *e.g.* for

- *input devices*, *e.g.* keyboard and mouse input.
- *prompts* with
  - *type–ahead*: several questions could be answered with one input, suppressing prompts which are known in advance.
  - *command abbreviation*: a command specifies the required program function; the user should only need to type enough first characters to distinguish the command from all others.
  - *line editing*: *e.g.* characters on the input line may be deleted.
  - *file input*: the sequence of commands and replies to prompts that the user would normally type at the keyboard, should also be readable from a file.
- *input validation* and error reporting: if invalid data are entered, or the program encounters any error, an explanatory message should be displayed, and the last prompt be repeated.

User Customisation Users often have different preferences (Linton *et al.*, 1989). For example, one user may prefer pop–up menus, while another prefers pull-down menus, while another prefers command line input. The user interface should provide a broad range of dialogue styles from which the users can select their preferences.

Screen and input management features such as those above should also be supported by character–oriented user interfaces; some may even support window management too (see TermWindows, below).

3.3 Architecture

A *dialogue* is the information interchange between the application program and the user, at the most general level. The information of interest is that which the user wants the program for, and not *e.g.* the instructions on how to display a table or graph. It comprises requests from the user, causing application actions and responses communicated back to the user. The dialogue is predicated on the existence of a repertoire of display capabilities maintained independently of the dialogue itself, which represents the essential components of the communication.

UNIX Graphical User Interfaces (GUI's) are typically based on the *client–server* model: a message handling interface encodes requests from the user interface to an appropriate protocol; and on the *application* side it decodes and parses, *i.e.* invokes dialogue functions. The user interface client—application server protocol in the dialogue layer can be modelled as a language (Robinson & Burns, 1985), executed on a virtual machine built from the objects referenced by the language, *e.g.* tables, forms and graphs. This allows the possibility of communicating between different computers; the user interface can then be distributed towards the end user, unloading centralised resources.

Applications such as STARCAT are usually developed together with their user interface. However, for remote users of these systems, it is very desirable to place the user interface as close to the user as possible, to reduce network overheads and to facilitate local customisation. The design should allow the user interface to be called directly, or via messages from a separate task (*e.g.* using Remote Procedure Calls (RPC)).

The window management system can be removed from the operating system (where Macintosh puts it), to be implemented as a user–level display server. Client applications, which may be distributed over a local area network (LAN), request graphics services from the server, by a byte stream protocol; the client views the server as an abstract device encapsulating display functions. Typically, window management is further separated out from the display, to a window server, which runs on a display device, and takes requests from a client application, and passes inputs from mouse, keyboard, *etc.* to the client.

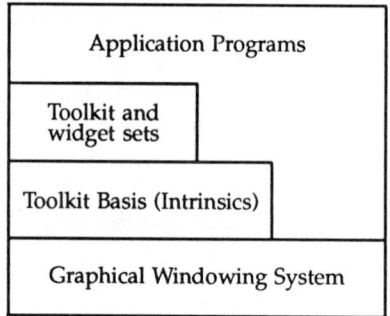

Figure 24–1: User Interface Architecture

This leads to a layered architecture (Mogg, 1989); see figure 24–1. Applications (perhaps built using a UIMS) may make use of any of the supporting layers, though the lower levels are harder to use—indeed this is one of the major *raisons d'être* for the higher levels. An Application Program Interface (API) provides access to user interface toolkits and widgets sets; this simplifies application programming and improves portability. A *widget* ('window gadget') is an input–output screen area, *e.g.* scroll bar, menu, box for closing, zooming, or resizing the window, *etc.*; widgets may be built on other widgets, and embellished with images, text, *etc.* This level may be built on a toolkit basis, *e.g.* in X, the X toolkit intrinsics. A Graphical Windowing System (GWS) provides the primitives for graphics operations and windowing, *e.g.* managing multiple simultaneous applications sharing a display device.

3.3.1 Control Location. Note that the user interface might well be layered on the application, not necessarily the other way round; user interfaces can be classified by how they communicate with the application programs at run–time (Myers, 1989):

Application control (also called internal control). The application calls interface procedures when input is needed. This method is used by some user interface toolkits.

User interface control (also called external control). The interface procedures call the application when the user gives a command. This method is used by most user interfaces because it lets the user interface handle scheduling and sequencing. It can be further classified by how the user interface communicates with the application:

Callback procedures The application passes to the user interface the names of procedures to call. The application is therefore organised as a set of procedures that the interface calls. This is probably the simplest and most efficient technique.

Shared memory The user interface and the application each poll the data to check for changes, or are automatically notified of changes.

Other methods are based on multiprocess message passing, or event handling mechanisms to communicate with applications. While they may be better structured, they are less efficient.

There may also be mixed control, where either the user interface or the application may assume control.

3.4 Design Models for User Interfaces

User Interface Management Systems have essentially been based on a few basic models of man–machine dialogue (Green, 1985, Myers, 1989). None of these models alone is ideal for all requirements, but the most popular has been the Context Free Grammar (CFG) model, for at least 2 reasons: (a)it is virtually trivial to implement, since the techniques are now thoroughly researched and documented, (see *e.g.* Aho *et al.*, 1986); and (b)it matches the concept of dialogue quite closely. In this paper we do not describe all classes of model, nor all examples of each; see Myers, 1989 for a more thorough treatment.

3.4.1 Toolkits. A library of interaction techniques (menus, graphical scroll bars, buttons, *etc.*) that can be called by the application. Using a library means that more time will become available for developing the application, and the user interface format will be similar across applications, saving the users time from re-learning. Libraries tend not to provide much support for dialogue control. Examples:

Macintosh Toolbox contains window, dialogue, menu, control, and textedit managers. These tools are accessible from a number of systems and languages, notably the Macintosh Programmer's Workshop (MPW). Their use requires a detailed knowledge of the ROM tools; a lot of effort is needed to master all the routines, which led Apple to create MACAPP.

Microsoft Windows for the IBM PC, is an application on MS-DOS. Multiple applications can be resident at once, control can be switched among them, and data can be moved between them. Presentation Manager runs on OS/2, and was derived from MS Windows.

X11 Toolkits: X Windows is a network-oriented, server-based windowing system developed at MIT and supported by most vendors of UNIX workstations. Applications use the Xlib toolkit and widgets.

> Andrew includes objects that contain text, bitmaps, spreadsheets, animation editors, *etc.* Its composition mechanism allows them to be embedded in multimedia documents.
>
> OSF-Motif is supported on over 100 hardware, and 38 operating system platforms. In 1988 the Open Software Foundation issued a request for technology for a window system and a style guide. The chosen windowing system was X11, with DEC's User Interface Language (UIL) application program interface (API), and features of the Hewlett-Packard/Microsoft's window managers. The style guide (Apple) promotes a consistent 'look and feel' across applications.
>
> Open-Look was designed by Sun Microsystems and AT&T, based on initial technology licensed from Xerox Corp. Its overriding objective is to meet the needs of evolving operating system and display device technology over the next decade, yet offer interoperability with currently prominent user interfaces (Mogg, 1989). The Xt+ toolkit is implemented on X.

Proteus/TermWindows is a C programmer's toolkit for rapid prototyping, database interfacing, translator construction, and user interface management. TermWindows is a Window Management System, which supports Proteus on a large variety of terminal types, using a TERMCAP database. It also provides a LaTeX-like document help display system; the same source files can be used for on-line help, and for a hardcopy from laser printers. With suitable macro definitions, screen definition files (SDF) can be processed through LaTeX to give hardcopy. The SDF feature means that much of the user interface can be specified in files, without hard–coding into the application; this facilitates rapid prototyping and updating.

These interfaces are portable (VMS, Unix, MS–DOS), helping the portability of the application itself, which is also insulated from changes in the underlying systems. Both Proteus and TermWindows are available in the public domain.

XVT The Extensible Virtual Toolkit, from the Advanced Programming Institute, Boulder, Co., USA, can be used to write portable programs that run under Microsoft Windows, OS/2 Presentation Manager, the Macintosh Toolbox, and X–Windows. The XVT library is sandwiched between the application and the target machine GUI. Development from one environment is ported to another by recompiling and re-linking the code with the application libraries. Note that XVT can only support the 'lowest common denominator' between GUI's.

3.4.2 State transition diagrams. The dialogue proceeds according to a set of nodes in a network; the nodes specify:

- *valid responses*,
- *new states*; arcs out of each state are labelled with the input token that will cause a
- *transition* to the state at the other end of the arc. The arcs may also be labelled with
- *application procedures* to be called, and output to be displayed. If the arcs have recursive calls to other diagrams, they are called
- *recursive transition networks* (RTN's).

User Software Engineering is a methodology for the specification and implementation of interactive systems. An early step in the methodology is the creation of a formal executable description of the user interaction with the system based on augmented state transition diagrams [27]. USE transition diagrams are based on perceived shortcomings of the pure state transition diagram approach.

3.4.3 Context–Free Grammar. The user–application dialogue can be viewed as comprising an exchange of tokens which are to be recognised and interpreted (parsed) according to a context–free grammar (Foley & van Dam, 1982). The dialogue spans 3 levels:

1. LEXICAL: input/output level; *e.g.* tokens as character strings such as command;

2. SYNTACTIC: recognition process, *e.g.* comparison of token against command table;
3. SEMANTIC: invocation process, *e.g.* action corresponding to user's command.

As a conceptual model, the layered architecture can be viewed as a series of higher-level languages being interpreted on lower-level virtual machines. Its major shortcoming is its orientation on a time-serial flow of tokens (Williams, 1986); parallelism and asynchronicity are not accounted for. The partitioning into lexical, syntactic and semantic phases complicates any continuous feedback mechanism, due to the discretisation into tokens. For example, if the user types a command and its parameters, then there is no simple way for the interface to report errors as they happen, such as a mis-spelling of the command (Kamran, 1985). However, it is cheap and easy to implement. A hybrid with object-oriented programming (Coutaz, 1987, Goldberg, 1986, Richmond, 1986, Sibert *et al.*, 1986) can combine their best characteristics.

Acknowledgements

We wish to gratefully acknowledge many fundamental discussions with François Ochsenbein, Benoît Pirenne and Guido Russo on theory and practice of user interfaces for astronomical applications. The authors are also grateful to Gianni Raffi and Mauro Comin for having provided information on the status and future developments of the ESO Remote Control interface. Faranguiss Poncet and Andy Götz, of the ESRF, provided invaluable information on the rapidly evolving state of UIMS's and toolkits, and kindly reviewed the last section of this chapter. M. Albrecht's help was particularly appreciated in defining the scope and tailoring the contents of this paper.

References

[1] Aho A.V., Sethi R. & Ullman J.D. (1986), *Compilers: Principles, Techniques, and Tools*. Addison-Wesley, Reading, Massachusetts.
[2] Bass L.J. (1985), "A generalized user interface for applications programs (2)", *Communications of the ACM,* **28**, 6.
[3] Bass L.J., & Bunker R.E. (1981), "A generalized user interface for applications programs", *Communications of the ACM,* **24**, 12.
[4] Coutaz J. (1985), "Abstractions for user interface design", *IEEE Computer,* **18**, 9.
[5] Coutaz J. (1987), "The construction of user interfaces and the object paradigm", in *Proceedings of the European Conference on Object-Oriented*

Programming, AFCET, Paris, pp. 135–144.
[6] Driessen C. & Pasian F. (1988), "USSP version 1.0 — User's Guide" *ESA IUE Newsletter*, 30, July 1988.
[7] Foley J.D. & van Dam A. (1982), *Fundamentals of Interactive Computer Graphics*. Addison–Wesley, 1982.
[8] Götz, A. (1990), "Evaluation of NASA's Transportable Application Environment", ESRF Technical Note.
[9] Goldberg A. (1986), "The influence of an object–oriented language on the programming environment", in *Interactive Programming Environments*, D.R. Barstow *et al.*(Eds.), McGraw Hill.
[10] Green M. (1985), "Report on dialogue specification tools", in *User Interface Management Systems*, G.Pfaff (Ed.), pp. 9–20, Springer–Verlag, Heidelberg.
[11] Kamran A. (1985), "Issues pertaining to the design of a user interface management system", in *User Interface Management Systems*, G.Pfaff (Ed.), pp. 43–48, Springer–Verlag, 1985.
[12] Linton M.A., Vlissides J.M. & Calder P.R. (1989), "Composing user interfaces with INTERVIEWS", *IEEE Computer*, 22, 2.
[13] McCartney J.E., Jefferys W.H. & McArthur B.E. (1990), "Pickles—A Graphical Tool for Pointing the Hubble Space Telescope and Displaying the Guide Star Catalog", Version 3.04 — January 1990.
[14] MIDAS Users Guide — Release 89NOV. Image Processing Group, ESO, November 1989
[15] Mogg, G. (1989), "The OPEN LOOK Graphical User Interface", *European UNIX systems User Group Newsletter*, 9, 4.
[16] Myers B.A. (1989), "User–interface tools: Introduction and survey", *IEEE Software*, 6, 1.
[17] Ochsenbein F. (1988), "TermWindows – A Terminal-Independent Screen Formatting and Windows Package", ESO/IPG, July 1988.
[18] Pirenne B., Hunt L., Richmond A., Russo G. (1988), "STARCAT — a complete database front–end", In *Astronet Workshop on Archives and Databases*, Bologna, Italy, December 1988.
[19] Pasian F. & Romelfanger F. (1989), "The operator's interface to the DMF", *ST-ECF technical report* **TR(ARC)-10**, August 1989.
[20] Pasian F. & Vuerli C. (1990), "DMF Operator Interface Process — User Guide.", ST-ECF technical report **TR(ARC)-13**, February 1990.
[21] Proteus Software Handbook. ST-ECF O-02 Document, Vol. VIII, Version 1.5, July 1988.
[22] Robinson J. & Burns A. (1985), "A dialogue development system for the design and implementation of user interfaces in Ada", *The Computer Journal*, 28, 1, February 1985.
[23] Richmond A. (1986), "Software design by object–oriented functional layering", *Computer Physics Communications*, 41, pp. 377–384, North–Holland, Amsterdam.
[24] Richmond A., McGlynn T., Ochsenbein F., Romelfanger F., Russo G. (1987), "The design of a large astronomical database system", in *Astronomy from Large Databases: Scientific Objectives and Methodological Approaches*,

ST–ECF/ESO, Garching, FRG, October 1987.
[25] Sibert J.L., Hurley W.D. & Bleser B.L. (1986), "An object–oriented user interface management system", *Computer Graphics*, **20**, 4, SIGGRAPH '86 Conference Proceedings.
[26] STARCAT User Guide. ST-ECF O-02 Document, Vol. VI, Version 2.25, July 1989.
[27] Wasserman A.I. (1985), "Extending state transition diagrams for the specification of human–computer interaction", *IEEE Transactions on Software Engineering*, **SE–11**, 8, pp. 699–713.
[28] Williams A. (1986), "An architecture for user interface R&D", *IEEE CG&A*, pp. 39–50, July 1986.
[29] Wamsteker W., Driessen C., Munoz J.R., Hassall B.J.M., Pasian F., Barylak M., Russo G., Egret D., Murray J., Talavera A., Heck A. (1989), "IUE-ULDA/USSP: The on-line low resolution spectral data archive of the International Ultraviolet Explorer", *Astron. Astrophys. Suppl. Series*, **79**, pp. 1-10.

25

The FITS Data Format

Preben Grosbøl
European Southern Observatory
Karl–Schwarzschild Straße 2
D–8046, Garching, Fed. Rep. of Germany

25–1 Introduction

The first proposal of a Flexible Image Transport System for astronomy was made by Wells and Greisen (1979). It described a general way to encode both a definition of the data and the data itself in a machine independent way using magnetic tape as the standard transport medium. The advantage of using a standard format for transport of astronomical images was soon realized and most major observatories implemented it as the prime format for data exchange. Subsequently, the FITS tape format (Wells *et al.*, 1981) was recommended as the standard format for interchange of image data between all observatories by Commission 5 at the 1982 General Assembly of International Astronomical Union (IAU) in Patras (*IAU Inf. Bull.* **49**, 1983). The first extension of the FITS format was also recommended during that meeting. This *"random-groups"* extension (Greisen and Harten, 1981) defined the way to transport a large number of data matrices with irregular spacing.

During the Patras meeting of IAU Commission 5, the possible extension of the FITS format to table and catalog data was discussed. As a result, a FITS Task Force under the Working Group on Astronomical Data was made to define and test a table extension to the FITS standard. The work of this Task Force resulted in a proposal for rules for generalized extensions to FITS (Grosbøl *et al.*, 1988) with a special extension for table and catalog data (Harten *et al.*, 1988). Further, a specification for the physical blocking of logical FITS data records of 2880 bytes was given to make more efficient use of high recording densities and new data media such as optical disks and helical scan cartridge tape devices. These extensions to FITS were recommended by the IAU General Assembly in Baltimore 1988 (*IAU Inf. Bull.* **61**, 1989) where also the creation of a permanent FITS Working Group was decided.

In the following section the general structure of a FITS file is given together with details of the different data formats provided. Finally, the organization of FITS standard committees is outlined describing how proposals for new extensions should be made to follow the accepted standard.

25-2 General Structure

A FITS file contains a sequence of logical units which all start with a set of header records describing the following data records. The logical record length of a FITS file is always 2880 bytes of 8 bits. Both header and data sections start in a new logical record. FITS headers are encoded in ASCII as 80 character card images each starting with an 8 character keyword defining the type of information contained on the card. Values of parameters are decoded using standard FORTRAN-77 rules. They describe in detail the data following the header records. After the last header/data unit in the file additional records may exist.

This design of the header gives a large flexibility to define the parameter necessary to describe a data set. By demanding that a well defined set of standard keywords is always given, the basic specification of the data can be read easily even by simple programs. The modular structure of a FITS file also makes it easy to transport more complex data. Further, it provides a possibility to extend the FITS standard to a new type of data (Grosbøl *et al.*, 1988) without modifying the present definition of FITS. In the following sections a number of such extensions are discussed.

25-3 The FITS Data Format

A large variety of data formats can be transported using FITS. The main emphasis has been on exchange of digital images although other types of data have recently been included. The different formats and data types are summarized below.

3.1 Prime data matrix

The first FITS paper by Wells *et al.* (1981) defined how a data matrix or multi-dimensional image sampled with uniformed spacing can be encoded. In the original agreement, only three data formats were supported namely, 8 bit unsigned, 16 and 32 bit signed twos-complement integers. The 32/64-bit IEEE floating point formats were added in January 1990 to accommodate data with a high dynamic range. Besides the definitions of the size and type of the data, a specification of the world (or physical) coordinates associated to the image can be given. A set of standard keywords is used to indicate associated parameters like source identification, observing data, origin and general comments.

3.2 Random groups format

Often sets of equal sized images with objects from different parts of the sky are obtained. It would be possible to write each individual image out in a FITS file. However, this creates a large overhead for small images due to repeated header information. The *"random groups"* extension to basic FITS was designed by Greisen and Harten (1981) to solve this problem. By using a special convention for the definition of the data matrix (i.e. increasing the dimension of the matrix by one and setting the size of the first axis to zero), a set of images with associated parameters could be encoded. The allowed data types are the same as for the prime data matrix.

3.3 Generalized extension

Although the flexibility of the header and image formats of the basic FITS definition provides wide possibilities to include other types of data, it was realized that only a well defined set of rules for future extensions of the FITS format could ensure the necessary stability of the standard. Such rules for generalized extensions to FITS was defined by Grosbøl et al. (1988). They open the possibility of having a sequence of header/data groups in a single FITS file. Each group starts with a header which has the same general format as the main header. The first card of the extension header defines the type of the following extension. Since the number of bits in the associated data records is specified, a reading program can skip over an unknown extension type to the next header/data group.

3.4 ASCII Table Extension

As the first extension to FITS following these rules, a table and catalog extension was defined by Harten et al. (1988). This extension makes it possible to exchange heterogeneous data which can be represented as a printable table. The data is given as a two dimensional matrix of ASCII characters with specifications of each column (or field). A given field must have a fixed format (following the FORTRAN-77 format rules) but can be of any type *e.g.* real, integer or character string. The direct mapping of a printed table or catalog makes the format easier to interpret and well suited for interchange.

3.5 Blocking of FITS

The basic FITS paper specified both a logical and physical record length of 2880 bytes. The increasing volume of data and higher recording densities made this physical record size inefficient. To increase storage efficiency and make use of new recording media such as optical disks and helical scan devices, the FITS standard was extended to allow physical blocking factors different from one (Grosbøl et al., 1988). The allowed range of blocking factors is explicitly defined for a given media. For normal 1/2-inch 9-track magnetic tapes, factors between 1 and 10 were allowed giving a maximum physical block length of 28800 bytes.

3.6 Other extensions under consideration

A number of possible extensions to the FITS standard have been under discussion in the FITS community. The three most important suggestions are currently: a) a high efficiency binary table format for e.g. X-ray detectors, b) blocking convention for FITS files on other media such as helical scan tapes, optical disk and tape cartridges, and c) extension of the world coordinate definitions.

25–4 FITS Committees and Standards

The FITS standard is controlled at an international level by a permanent IAU FITS Working Group which was created at the 1988 IAU General Assembly in Baltimore (*IAU Inf. Bull.* **61**, 1989). This Working Group has representatives from major astronomical institutes and data centers. For general discussions of the FITS standard and possible new extensions a number of regional committees have been formed e.g. the North American FITS group under the Working Group for Astronomical Software of American Astronomical Society, the European FITS Committee and the Japanese FITS Group. A FITS support office has been established by NASA as a part of NASA Science Data Systems Standard Office at the Goddard Space Flight Center.

The normal way for a proposal of a new FITS extension is first a general discussion in the regional groups. If a given proposal is found necessary because it provides efficient encoding of new types of data which cannot be done with the existing formats, these groups should pass motions recommending it *after* test implementations have verified the proposed extension format. When all regional FITS committees agree on a proposal, the IAU FITS Working Group will make a final standards document which will be passed to a reviewing body representing major institutes and data centers. If the final proposal is accepted in the review, it will be proposed as a FITS standard to be recommended by IAU.

25-5 Conclusions

The concept of the FITS format has proven extremely useful for astronomers who can now exchange data between different institutes and/or image processing systems just by specifying that they should be written in FITS format. The FITS committees ensure that this standard can be maintained in the future. The adoption of the rules for generalized extensions of FITS provides a well defined method with which new needs can be accommodated without changes in the present standard.

References

[1] Greisen E. W. & Harten R. H. (1981), *Astron. Astrophys. Suppl.*, 44, 371.
[2] Grosbøl P., Harten R. H., Greisen E. W., Wells D. C. (1988), *Astron. Astrophys. Suppl.*, 73, 359.
[3] Harten R. H., Grosbøl P., Greisen E. W., Wells D. C. (1988), *Astron. Astrophys. Suppl.*, 73, 365.
[4] *IAU Inf. Bull.* 49, 14 (1983).
[5] *IAU Inf. Bull.* 61, 10 (1989).
[6] Wells D. C. & Greisen E. W. (1979), 'FITS: a Flexible Image Transport System' in *Image Processing in Astronomy*, G. Sedmak, M. Capaccioli, R.J. Allen (Eds.), Trieste, 445.
[7] Wells D. C., Greisen E. W., Harten R. H. (1981), *Astron. Astrophys. Suppl.*, 44, 363.

Acronyms

AAS	American Astronomical Society
ADC	Astronomical Data Center
ADS	Astrophysics Data System
AG	Astronomische Gesellschaft
AIPS	Astronomical Image Processing System
AIT	Astronomisches Institut der Universität Tübingen
ANSA	Advanced Network System Architecture
API	Application Program Interface
ARI	Astronomisches Rechen Institut
ASpScROW	Astronomy, Space Sciences and Related Organizations of the World
AURA	Association of Universities for Research in Astronomy
AXAF	Advanced X-ray Astronomy Facility
BD	Bonner Durchmusterung
BITNET	Because It's Time Network
BRS	Bibliographic Retrieval Service
BSI	Bibliographic Star Index
CADC	Canadian Astronomy Data Centre
CCDM	Catalogue des Composantes d'étoiles Doubles et Multiples
CCSDS	Consultative Committee for Space Data Systems
CD-ROM	Compact Disk – Read Only Memory
CDS	Centre de Données astronomiques de Strasbourg
CERN	Centre Européen de Recherche Nucléaire
CFG	Context Free Grammar
CFHT	Canada-France-Hawaii Telescope
CNES	Centre National d'Etudes Spatiales
COBE	Cosmic Background Explorer
COSINE	Cooperation for Open Systems Interconnection Networking in Europe
CPI	Conference Papers Index
CRDD	Calibrated Raw Detector Data
CSI	Catalog of Stellar Identifications

DADS	Data Archive and Distribution System
DAO	Dominion Astrophysical Observatory
DARPA	Defense Applied Research Projects Agency
DAVID	Distributed Access View Integrated Database
DBIMA	DataBase of Images
DBMS	Database Management System
DCL	Digital Command Language
DDBMS	Distributed Database Management System
DFN	Deutsches Forschungsnetz
DMF	Data Management Facility
EARN	European Academic and Research Network
ECMWF	European Centre for Medium-Range Weather Forecasting
EFOSC	ESO Faint Object Spectrograph and Camera
EMMI	ESO MultiMode Instrument
ESA	European Space Agency
ESIS	European Space Information System
ESO	European Southern Observatory
E-SPAN	European SPAN
ESRIN	European Space Research Institute
FITS	Flexible Image Transport System
FIZ	FachInformationZentrum
FK5	Fifth Fundamental Catalogue
FUSE	Far Ultraviolet Spectroscopic Explorer
GARR	Gruppo Armonizzazione delle Reti per la Ricerca
GO	Guest Observers
GRO	Gamma-Ray Observer
GSC	Guide Star Catalog
GSFC	Goddard Space Flight Center
GSOC	German Space Operation Center
GSSS	Guide Star Selection System
GUI's	Graphical User Interfaces
GWS	Graphical Windowing System
HDS	Hierarchical Data System
HEA-SARC	High Energy Astrophysics Science Archive Research Center
HEPnet	High Energy Physics network
HIPPARCOS	High Precision PARallax COllecting Satellite
HST	Hubble Space Telescope
IAA	International Aerospace Abstracts
IAC	Instituto Astrofisica de Canarias

IAP	Institut d'Astrophysique de Paris
IAU	International Astronomical Union
IDAAS	International Directory of Astronomical Associations and Societies
IDI	Images Display Interfaces
IDPAI	International Directory of Professional Astronomical Institutions
IEEE	Institute of Electrical and Electronics Engineers
INCA	Input Catalogue (Hipparcos)
ING	Isaac Newton Group of Telescopes
INSPEC	Information Services in Physics, Electrotechnology, Computers and Control
INSU	Institut National des Sciences de l'Univers
IPAC	Infrared Processing and Analysis Center
IPMAF	IRAS Post Mission Analysis Facility
IRAF	Image Reduction and Analysis Facility
IRAS	InfraRed Astronomical Satellite
ISDN	Integrated Services Digital Network
ISI	Institute for Scientific Information
ISO	International Standards Organization
IUE	International Ultraviolet Explorer
IXI	International X.25 Interconnect
JANET	Joint Academic NETwork
JCMT	James Clerk Maxwell Telescope
JPL	Jet Propulsion Laboratory
LAN	local area network
LRS	Low Resolution Spectrometer
LRS	Low Resolution Spectrometer
MD	Master Directory
MFEnet	Magnetic Fusion Energy network
MIDAS	Munich Image Data Analysis System
MIT	Massachusetts Institute of Technology
MPE	Max Planck Institut für Extraterrestrische Physik
NAC	Nederlandse Astronomenclub
NED	NASA/IPAC Extragalactic Database
NFRA	Netherlands Foundation for Research in Astronomy
NIVR	Netherlands Agency for Aerospace Programmes
NRAO	National Radio Astronomy Observatory
NREN	National Research and Education Network
NSDSSO	NASA Science Data Systems Standard Office
NSF	National Science Foundation
NSI	NASA Science Internet

NSN	NASA Science Network
NSSDC	National Space Science Data Center
NTIS	National Technical Information Service
NTT	New Technology Telescope
OAO-3	Orbiting Astronomical Satellite 3
OOP	Object Oriented Programming
OSI	Open Systems Interconnect
OSSA	Office of Space Science and Applications
PAD	Packet Assembler-Deassembler
PEPSI	Proposal Entry Processing System
PI	Principal Investigator
QBE	Query by Example
RAL	Rutherford Appleton Laboratory
RAL SDC	RAL Space Data Centre
RARE	Réseaux Associés pour la Recherche Européenne
RGO	Royal Greenwich Observatory
ROB	Royal Observatory of Belgium
ROE	Royal Observatory Edinburgh
ROSAT	Röntgen Satellite
RPC	Remote Procedure Calls
RSCS	Remote Spooling Communication Subsystem
RSDC	ROSAT Scientific Data Center
SAO	Smithsonian Astrophysical Observatory
SCAR	Starlink Catalogue Access and Reporting
SERC	Science and Engineering Research Council
SIG	Special Interest Group
SIMBAD	Set of Identifications, Measurements, and Bibliography for Astronomical Data
SIRTF	Space Infra-Red Telescope Facility
SMM	Solar Maximum Mission
SPAN	Space Physics Analysis Network
SQL	Structured Query Language
ST–ECF	Space Telescope–European Coordinating Facility
STADAT	Starlink database
STARCAT	Space Telescope Archive and Catalogue
STAR	NASA Scientific and Technical Aerospace Reports
STN	Scientific and Technical Information Network
STScI	Space Telescope Science Institute
STSDAS	Space Telescope Science Data Analysis System
TAE+	Transportable Application Environment

TCP/IP	Transfer Control Protocol/Interaction Protocol
UIL	User Interface Language
UIMS	User Interface Management System
UKIRT	United Kingdom Infrared Telescope
ULDA	Uniform Low Dispersion Archive
ULP	Université Louis Pasteur, Strasbourg
USSP	ULDA Support Software Package
VLA	Very Large Array
VLBA	Very Long Baseline Array
VLT	Very Large Telescope
WDC-A R&S	World Data Center-A for Rockets and Satellites
WFC	ROSAT Wide Field Camera
WIN	Wissenschaftsnetz
WORM	Write Once, Read Many times
WSDB	[IRAS] Working Survey DataBase
WSRT	Westerbork Synthesis Radio Telescope

Index

8mm tape   157
9–track tape   157
AAS   220
abstract   99
access mode   8, 42, 75, 84, 99, 119
acronym   215
ADAM   27, 164, 168
ADC   152
ADS   85, 87, 139, 156, 198, 199
Advanced Network System Architecture see ANSA
Advanced X-ray Astronomy Facility see AXAF
Aerospace Database   200
AG   220
AIPS   118, 125, 238
AIT   3
American Astronomical Society see AAS
analysis software   5, 149
analysis system   14
ancillary data   117, 118
ANSA   143
API   246
application procedure   249
Application Program Interface see API
archival research   12
archive   38
ARI   86
Ariane launcher   67
ARPAnet   226
artificial intelligence   24
ASpScROW   **211**, 214
Association of Universities for Research in Astronomy see AURA
astrometric data   81
astrometry   61

ASTRONET   **179**, 190
Astronews   56
astronomical catalogue   4, 19, 56, 72, 89, 93, 113, 152, 164, 166, 179
Astronomical Data Center see ADC
astronomical image   189
Astronomical Image Processing System see AIPS
astronomical object   63, 182
Astronomische Gesellschaft see AG
Astronomisches Institut der Universität Tübingen see AIT
Astronomisches Rechen Institut see ARI
Astronomy and Astrophysics Abstracts   200
Astronomy, Space Sciences and Related Organizations of the World see ASpScROW
Astrophysics Data System see ADS
AURA   59
AXAF   90

basic data   94
Bayesian classifier   61
BD   152
Because It's Time Network see BITNET
Bibliographic Retrieval Service see BRS
Bibliographic Star Index see BSI
bibliographical reference   80, 81, 86, 96, **199**
BITNET   85, 151, 225, 226
Bonner Durchmusterung see BD
BRS   208
BSI   80

C language 50, 75, 82, 111
CA*net 196
CADC 53, 113, **193**
calibrated data 107
Calibrated Raw Detector Data see CRDD
Calibration Database 195
Canada-France-Hawaii Telescope see CFHT
Canadian Astronomy Data Centre see CADC
Catalog of Stellar Identifications see CSI
Catalogue des Composantes d'étoiles Doubles et Multiples see CCDM
catalogue name 183
CCDM 73
CCSDS 162
CD-ROM 60, 62, 76, 152, 164, 182, 195
CDS 20, 42, 75, 76, 79, 113, 133, 152, 164, 179
Centre de Données astronomiques de Strasbourg see CDS
Centre Européen de Recherche Nucléaire see CERN
Centre National d'Etudes Spatiales see CNES
CERN 230
CFG 247
CFHT 197
class 83
classification 24, 61, 64, 92, 98, 198
CLUSTER 164
CNES 75, 87, 228
COBE **29**, 154
command 83, 136
command abbreviation 245
Compact Disk – Read Only Memory see CD-ROM
compilation catalogue 166
CONF 201
conference 201
Conference Papers Index see CPI
Consultative Committee for Space Data Systems see CCSDS
context 145
Context Free Grammar see CFG
context help 120
contextual heterogeneity 132
Cooperation for Open Systems Interconnection Networking in Europe see COSINE
coordinate system 33, 100, 174
coordinate transformation 99
coordinates 80, 113, 119, 183, 211
COSINE 230
Cosmic Background Explorer see COBE
cosmos 90, 167
CPI 201
CRDD 163
cross correlation 137
cross identification 64, 69, 80, 92
CRRES 164
CSI 80

DADS 48, 198
DAO 37, 193, 194
DARPA 226
data analysis package 42
Data Archive and Distribution System see DADS
data descriptor 107
data documentation 155
Data Management Facility see DMF
data model 131
data rights 8, 11, 30, 36, 41, 51, 108, 117, 118, 125, 147, 154
Database Management System see DBMS
DataBase of Images see DBIMA
database vendor 208
DATEX-P 230
DAVID 144, 156
DBIMA 179, 189
DBMS 6, 13, 82, 104, 112, 131, 144
DCL 180
DDBMS 156
DECnet 119

Index 267

Defense Applied Research Projects Agency *see* DARPA
Deutsches Forschungsnetz *see* DFN
DFN 230
Digital Command Language *see* DCL
DIRA 179
directory 129, 140, **211**, 220, 230
Distributed Access View Integrated Database *see* DAVID
Distributed Database Management System *see* DDBMS
distributed system 146
DMF 48, 193
documentation 149, 183
Dominion Astrophysical Observatory *see* DAO
double stars 71

EARN 225, 226
ECMWF 230
EFOSC 111
Einstein observatory 2, 13, 198
electronic mail 100, 102, 119, 129, 215, 219, 221, 226, 230
EMMI 109
entity–relationship diagram 133, 136
equinox 113
ERS-1 164
ESA 11, 35, 48, 67, 75, 76, 85, 130, 133, 220, 226, 230
  IRS 129
ESA Headquarters 228
ESANET 129, 162
ESIS 11, 85, 86, **127**, 162, 199, 223
ESO 5, 56, **107**, 220
ESO Faint Object Spectrograph and Camera *see* EFOSC
ESO MultiMode Instrument *see* EMMI
ESOC 228
E-SPAN 228
ESRIN 209, 228
essential note 96
ESTEC 13, 228

European Academic and Research Network *see* EARN
European Centre for Medium-Range Weather Forecasting *see* ECMWF
European Southern Observatory *see* ESO
European Space Agency *see* ESA
European Space Information System *see* ESIS
European Space Research Institute *see* ESRIN
European SPAN *see* E-SPAN
Exabyte 162
EXOSAT 5, **11**, 38, 129, 199, 238
expert system 156
Extragalactic database **89**

FachInformationZentrum *see* FIZ
facsimile 232
Factor Space Access Method 145
Far Ultraviolet Spectroscopic Explorer *see* FUSE
Fifth Fundamental Catalogue *see* FK5
file input 245
file transfer 226
filter 81
FITS 5, 17, 23, 30, 34, 62, 109, 116, 117, 121, 137, 153, 155, 164, 190, **253**
  header 62
  table 62, 63, 253
FIZ 205
FK5 152
Flexible Image Transport System *see* FITS
flux 81
format 17, 42, 81, 137, 253
Fortran 180
Fortran 77 171
FREYJA 164
FUSE 156

Gamma-Ray Observer *see* GRO
GARR 230
gateway 225

GEISHA 26
Geophysical Data Facility 129
German Space Operation Center *see*
    GSOC
GO 36
Goddard Space Flight Center *see*
    GSFC
graphic capabilities 238
graphic task 187
graphical metafiles 137
Graphical User Interfaces *see* GUI's
Graphical Windowing System *see*
    GWS
graphics 13
GRO 90, 154, 156
groundbased observatory 115
Gruppo Armonizzazione delle Reti
    per la Ricerca *see* GARR
GSC 13, 56, **59**, 86, 149, 195, 242
GSFC 3, 30, 36, 37, 48, 140
GSOC 3
GSSS 48
Guest Investigator program 31
Guest Observers *see* GO
GUI's 245
Guide Star Catalog *see* GSC
Guide Star Selection System *see*
    GSSS
GWS 247

HDS 27, 175
HEA-SARC 157
header 254
HEPnet 228
Hierarchical Data System *see* HDS
High Energy Astrophysics Science
    Archive Research Center *see*
    HEA-SARC
High Energy Physics network *see*
    HEPnet
High Precision PARallax COllecting
    Satellite *see* HIPPARCOS
HIPPARCOS **67**, 79
    Input Catalogue **67**
HST 38, **47**, 59, 90, 193
Hubble Space Telescope *see* HST

IAA 203, 204
IAC 116
IAP 86
IAU 109, 221, 253
IDAAS 214
identification 63, 80
identifier 68, 73
IDI 191
IDL 164, 238
IDPAI 214
IEEE 255
IHAP 238
image processing 110
Image Reduction and Analysis
    Facility *see* IRAF
Images Display Interfaces *see* IDI
INCA **67**, 79
Information Services in Physics,
    Electrotechnology,
    Computers and Control *see*
    INSPEC
InfraRed Astronomical Satellite *see*
    IRAS
Infrared Processing and Analysis
    Center *see* IPAC
ING **115**, 116
INGRES 6
Input Catalogue (Hipparcos) *see*
    INCA
input device 245
input validation 245
INSPEC 200, 203, 205
Institut d'Astrophysique de Paris
    see IAP
Institut National des Sciences de
    l'Univers *see* INSU
Institute for Scientific Information
    see ISI
Institute of Electrical and Electronics
    Engineers *see* IEEE
Instituto Astrofisica de Canarias *see*
    IAC
INSU 79, 87
Integrated Services Digital Network
    see ISDN
International Aerospace Abstracts

Index 269

see IAA
International Astronomical Union
  see IAU
International Directory of
  Astronomical Associations
  and Societies see IDAAS
International Directory of
  Professional Astronomical
  Institutions see IDPAI
International Standards
  Organization see ISO
International Ultraviolet Explorer
  see IUE
International X.25 Interconnect see
  IXI
Internet 11, 85, 101, 112, 151, 196,
  226, 229
IPAC 20, 24, 89, 140, 163, 164
IPMAF 24, 168
IRAF 164, 195, 238
IRAS 17, 149, 151, 156, 163, 167,
  195, 199
IRAS Point Source catalogue 22
IRAS Post Mission Analysis Facility
  see IPMAF
[IRAS] Working Survey DataBase
  see WSDB
Isaac Newton Group of Telescopes
  see ING
ISDN 232
ISI 202, 206
ISO 26, 27, 90, 142, 164, 229
  OSI 226
Itapac 230
IUE 35, 48, 108, 129, 147, 149, 151,
  156, 163, 166, 195, 199, 237
  archive 36
  Merged Log 39, 41
  observatory 36
  ULDA 13
IXI 230

James Clerk Maxwell Telescope see
  JCMT
JANET 164, 225, 228, 230
JCMT 119, 123

Jet Propulsion Laboratory see JPL
Joint Academic NETwork see
  JANET
JPL 17

Kerberos 147
keyword 109, 154, 220
kinematic data 98
Knowledge Dictionary System 143

La Palma Observatory 115, 166
LAN 246
lexical analysis 249
light-curve 12
line editing 245
local area network see LAN
Low Resolution Spectrometer see
  LRS
LRS 22, 24, 163
Lyman 197

magnetic disk 189
Magnetic Fusion Energy network
  see MFEnet
magnitude 64, 70, 80, 81
Massachusetts Institute of
  Technology see MIT
Master Directory see MD
Max Planck Institut für
  Extraterrestrische Physik
  see MPE
MD 153
measurement 81
message transmission 83
method 83
MFEnet 228
MIDAS 5, 108, 110, 164, 238
MIT 147
MPE 3
MS-DOS 125
multiple system 71
Munich Image Data Analysis System
  see MIDAS

NAC 221
NASA 27, 35, 48, 139, 151, 200, 226,
  244

IPAC Extragalactic Database see NED
RECON 203
NASA Ames Research Center 17
NASA Headquarters 31
NASA Science Data Systems Standard Office see NSDSSO
NASA Science Internet see NSI
NASA Science Network see NSN
NASA Scientific and Technical Aerospace Reports see STAR
National Radio Astronomy Observatory see NRAO
National Research and Education Network see NREN
National Science Foundation see NSF
National Space Science Data Center see NSSDC
National Technical Information Service see NTIS
natural language 145
NED 86, **89**, 147
Nederlandse Astronomenclub see NAC
Netherlands Agency for Aerospace Programmes see NIVR
Netherlands Foundation for Research in Astronomy see NFRA
Netherlands organization for scientific research 124
network 85, 101, 128, 129, 141, 151, 196, 221, **225**
network protocol 55, 225
New Technology Telescope see NTT
NFRA 116
NIVR 27
nomenclature 103
NRAO 118, 125
NREN 229
NSDSSO 155
NSF 226
NSFnet 196, 226

NSI 139, 141, 153, 228
NSN 142, 226, 228
NSSDC 16, 20, 30, 36, 113, 140, **151**, 179
NTIS 207
NTT 108, 242
null values 171
Numeris 232

OAO-3 156
object 83
object directory 92
object identifier 109
Object Oriented Programming see OOP
object type 93
observation
    catalogue 117, 166
    log 99, 109, 111
    run 110
Observatoire de Paris 68, 87
observatory 211
observing programme 68
Office of Space Science and Applications see OSSA
OOP 75, 82
open architecture 128
Open Systems Interconnect see OSI
optical disk 6, 42, 48, 51, 53, 62, 117, 118, 124, 151, 157, 162, 164, 182, 189, 253
Orbiting Astronomical Satellite 3 see OAO-3
OSI 51, 196, 229
OSSA 48

Packet Assembler-Deassembler see PAD
PAD 196
parallax 81
peculiar star 70
PEPSI 53
personal database 185
photometric data 98
photometry 61
physical heterogeneity 131
PI 54

Polar Platform 164
portability 82, 111
position 33, 64, 94
postscript 13, 238
Principal Investigator see PI
projection
  Aitoff 174
  tangent plane 174
prompts 245
proper motion 60, 70
Proposal Entry Processing System see PEPSI
Proteus 55, 241
public network 11, 112, 151, 196, 229
public switched networks 226

QBE 143, 149
quality control 87
Queen Mary and Westfield College 24
query 81, 120, 133, 136, 145, 153
Query by Example see QBE
query context 84
Quick-look 132

R-EXEC 164
radial velocities 81
RAL 3, 20, 36, 41, 133, 161, 165, 168
RAL SDC **161**
RAL Space Data Centre see RAL SDC
RARE 230
raw data 38, 107
recursive transition networks 249
reduced data 107
region 63
relational database 7, 20, 54, 83, 91, 104, 142, 169
Remote Procedure Calls see RPC
Remote Spooling Communication Subsystem see RSCS
RGO 116
ROB 211
ROE 167, 168
ROSAT **1**, 154, 198

ROSAT Scientific Data Center see RSDC
ROSAT Wide Field Camera see WFC
Royal Greenwich Observatory see RGO
Royal Observatory Edinburgh see ROE
Royal Observatory of Belgium see ROB
RPC 246
RSCS 226
RSDC 3
Rutherford Appleton Laboratory see RAL
Röntgen Satellite see ROSAT
Réseaux Associés pour la Recherche Européenne see RARE
Réunir 230

SAO 85, 140, 149, 152, 242
scanned Schmidt plates 166
SCAR 18, 123, 163, 164, 167, **168**
Science and Engineering Research Council see SERC
Scientific and Technical Information Network see STN
SCISEARCH 200
security 51, 147
self-documentation 102
semantic analysis 250
semantic heterogeneity 131
semantic network 133
SERC 27, 35, 124, 161, 165
session log file 120
Set of Identifications, Measurements, and Bibliography for Astronomical Data see SIMBAD
Sheffield University 133
SIG 179
SIMBAD 68, **79**, 91, 93, 97, 111, 129, 179, 195, 199, 205, 223
SIRTF 26, 90
sky map 9, 32, 34, 181
Sky & Telescope 221

Smithsonian Astrophysical
    Observatory *see* SAO
SMM   163
SOHO   164
Solar Maximum Mission *see* SMM
SOLAR-A   164
source identification   13, 87
Space Infra-Red Telescope Facility
    see SIRTF
Space Physics Analysis Network *see*
    SPAN
Space Research Institute in
    Groningen   24
Space Telescope Archive and
    Catalogue *see* STARCAT
Space Telescope Science Data
    Analysis System *see*
    STSDAS
Space Telescope Science Institute *see*
    STScI
Space Telescope–European
    Coordinating Facility *see*
    ST–ECF
SPAN   11, 75, 85, 101, 112, 141, 151,
    164, 221, 225, 226
spatial data   98
Special Interest Group *see* SIG
spectral classification   81
spectral type   70
SPECTRUM-X   164
SQL   41, 135, 143, 144, 149
ST–ECF   48, 112, 129, 193
STADAT   164
standards   155
STAR   203
Starcat   55, **112**, 195, 241
STARLINK   8, 19, 20, 26, 41, 123,
    161, **165**
Starlink Catalogue Access and
    Reporting *see* SCAR
Starlink database *see* STADAT
statistical research   123
STN   209
STN PHYS   200
storage media   6, 37, 50, 53, 110,
    154, 157, 162, 164, 256

Structured Query Language *see* SQL
STScI   47, 59, 112, 113, 140, 193
STSDAS   195, 238
subject matter   145
survey   1, 8, 18, 65, 69, 90
syntactic analysis   250
syntactic heterogeneity   131

TAE+   244
TCP/IP   142, 196, 226, 229
TermWindows   56, 187, 241
text management   142
time variability   4, 11
time-ordered data   31
Transfer Control
    Protocol/Interaction
    Protocol *see* TCP/IP
Transpac   230
Transportable Application
    Environment *see* TAE+
trigonometric parallax   70
type–ahead   245

UIL   248
UIMS   243
UKIRT   115, 119, 123
ULDA   42, 163, 195, 237
ULDA Support Software Package
    see USSP
ULP   79
Uniform Low Dispersion Archive
    see ULDA
United Kingdom Infrared Telescope
    see UKIRT
University College London   168
University of Colorado   37, 140
University of Leicester   8, 16
University of Leiden   20, 24
Université Louis Pasteur, Strasbourg
    see ULP
Université Paris XI   69, 87
UNIX   5, 56, 75, 84, 162, 190, 225
user interface   13, 52, 55, 81, 87, 101,
    112, 130, 142, 188, **235**
User Interface Language *see* UIL
User Interface Management System
    see UIMS

USSP   42, 237
UV–FITS format   118

variability   107
VAX/VMS   5, 26, 56, 117, 165, 180, 190, 226
Very Large Array *see* VLA
Very Large Telescope *see* VLT
Very Long Baseline Array *see* VLBA
VILSPA   36
VLA   **125**
VLBA   125
VLT   108
VMS   57

WDC-A R&S   152
Westerbok Observatory   **115**
Westerbork Synthesis Radio Telescope *see* WSRT
WFC   163
WIN   230
Wissenschaftsnetz *see* WIN
world coordinates   255
World Data Center-A for Rockets and Satellites *see* WDC-A R&S
WORM   110, 157, 162, 164, 182, 190
Write Once, Read Many times *see* WORM
WSDB   19
WSRT   115

X.25   112, 229
X.400   230

zodiacal light   23